Databases for Estimating HEALTH INSURANCE COVERAGE FOR CHILDREN

A Workshop Summary

Thomas J. Plewes, *Rapporteur*

Committee on National Statistics

Division of Behavioral and Social Sciences and Education

NATIONAL RESEARCH COUNCIL
OF THE NATIONAL ACADEMIES

THE NATIONAL ACADEMIES PRESS
Washington, D.C.
www.nap.edu

THE NATIONAL ACADEMIES PRESS 500 Fifth Street, N.W. Washington, DC 20001

NOTICE: The project that is the subject of this report was approved by the Governing Board of the National Research Council, whose members are drawn from the councils of the National Academy of Sciences, the National Academy of Engineering, and the Institute of Medicine. The members of the Steering Committee for the workshop were chosen for their special competences and with regard for appropriate balance.

This study was supported by Contract No. HHSP23320042509XI between the Office of the Assistant Secretary for Planning and Evaluation, Department of Health and Human Services, and the National Academy of Sciences. Support of the work of the Committee on National Statistics is provided by a consortium of federal agencies through a grant from the National Science Foundation (No. SES-0453930). Any opinions, findings, conclusions, or recommendations expressed in this publication are those of the author(s) and do not necessarily reflect the views of the organizations or agencies that provided support for the project.

International Standard Book Number-13: 978-0-309-16240-1
International Standard Book Number-10: 0-309-16240-8

Additional copies of this report are available from the National Academies Press, 500 Fifth Street, N.W., Lockbox 285, Washington, DC 20055; (800) 624-6242 or (202) 334-3313 (in the Washington metropolitan area); Internet, http://www.nap.edu.

Printed in the United States of America

Suggested citation: National Research Council. (2010). *Databases for Estimating Health Insurance Coverage for Children: A Workshop Summary*. Thomas J. Plewes, Rapporteur. Committee on National Statistics, Division of Behavioral and Social Sciences and Education. Washington, DC: The National Academies Press.

THE NATIONAL ACADEMIES

Advisers to the Nation on Science, Engineering, and Medicine

The **National Academy of Sciences** is a private, nonprofit, self-perpetuating society of distinguished scholars engaged in scientific and engineering research, dedicated to the furtherance of science and technology and to their use for the general welfare. Upon the authority of the charter granted to it by the Congress in 1863, the Academy has a mandate that requires it to advise the federal government on scientific and technical matters. Dr. Ralph J. Cicerone is president of the National Academy of Sciences.

The **National Academy of Engineering** was established in 1964, under the charter of the National Academy of Sciences, as a parallel organization of outstanding engineers. It is autonomous in its administration and in the selection of its members, sharing with the National Academy of Sciences the responsibility for advising the federal government. The National Academy of Engineering also sponsors engineering programs aimed at meeting national needs, encourages education and research, and recognizes the superior achievements of engineers. Dr. Charles M. Vest is president of the National Academy of Engineering.

The **Institute of Medicine** was established in 1970 by the National Academy of Sciences to secure the services of eminent members of appropriate professions in the examination of policy matters pertaining to the health of the public. The Institute acts under the responsibility given to the National Academy of Sciences by its congressional charter to be an adviser to the federal government and, upon its own initiative, to identify issues of medical care, research, and education. Dr. Harvey V. Fineberg is president of the Institute of Medicine.

The **National Research Council** was organized by the National Academy of Sciences in 1916 to associate the broad community of science and technology with the Academy's purposes of furthering knowledge and advising the federal government. Functioning in accordance with general policies determined by the Academy, the Council has become the principal operating agency of both the National Academy of Sciences and the National Academy of Engineering in providing services to the government, the public, and the scientific and engineering communities. The Council is administered jointly by both Academies and the Institute of Medicine. Dr. Ralph J. Cicerone and Dr. Charles M. Vest are chair and vice chair, respectively, of the National Research Council.

www.national-academies.org

STEERING COMMITTEE FOR A WORKSHOP ON EVALUATING DATABASES FOR USE IN THE CHILDREN'S HEALTH INSURANCE PROGRAM (CHIP) ALLOCATION FORMULA

V. JOSEPH HOTZ (*Chair*), Department of Economics, Duke University
LYNN A. BLEWETT, School of Public Health, University of Minnesota, and State Health Access Data Assistance Center
MICHAEL DAVERN, National Opinion Research Center and University of Chicago
LISA C. DUBAY, Bloomberg School of Public Health, Johns Hopkins University
JOAN HENNEBERRY, Colorado Department of Health Care Policy and Financing, Denver
CHRIS L. PETERSON, Medicaid and CHIP Payment and Access Commission, Washington, DC
ERIC V. SLUD, Department of Mathematics, University of Maryland

THOMAS J. PLEWES, *Study Director*
ANTHONY S. MANN, *Program Associate*

COMMITTEE ON NATIONAL STATISTICS
2010

Preface

This report summarizes the proceedings of a workshop convened in June 2010 to critically examine the various databases that could provide national and state-level estimates of low-income uninsured children and could be effectively used as criteria for monitoring children's health insurance coverage. The workshop was sponsored by the Office of the Assistant Secretary for Planning and Evaluation (ASPE), U.S. Department of Health and Human Services, and convened by the Committee on National Statistics (CNSTAT), Division of Behavioral and Social Sciences and Education (DBASSE) of the National Research Council (NRC).

The workshop was organized by a steering committee composed of experts in administering and assessing the Children's Health Insurance Program (CHIP) at the national and state levels, as well as experts in survey methodology and modeling. The committee provided invaluable guidance in developing the workshop, securing expert presentations, and facilitating the conduct of the workshop. Although the steering committee played a central role in designing and conducting the workshop, the members did not actively participate in writing this workshop summary.

The staff of ASPE and of the U.S. Census Bureau played an important role in preparing for and conducting the workshop. The work of Don Oellerich of ASPE and David Johnson of the Census Bureau in this preparation phase is especially recognized, as are the important information and comments they provided that enhanced the value of the event. As the person responsible for supervising the grant that supported this activity, Oellerich also served as the primary point of coordination between the steering committee and the Department of Health and Human Services.

The presentations in the workshop were designed to shed light on various aspects of the issues involved in evaluating the adequacy and appropriateness of databases for estimating health insurance coverage for children. The task of describing the overall context for the consideration of these issues fell to Chris Peterson of the Medicaid and CHIP Payment and Access Commission, a member of the steering committee. He discussed the evolution of the CHIP legislation and the more recent change in emphasis from use of coverage estimates for allocating funds to states to the purpose of assessing progress and performance overall that came with the reauthorization of CHIP in 2009. He described how the new health care reform legislation, known as the Patient Protection and Affordable Care Act, sets a new standard for the coverage data. Michael Davern, also a member of the steering committee, discussed the background of the major surveys that have been used to measure the health insurance coverage of children.

The first panel discussion focused on the use of administrative data for estimating the coverage of health insurance for children. David Baugh, who has responsibility for the Medicaid and CHIP databases at the Centers for Medicare & Medicaid Services (CMS), discussed the status of the major administrative data collections and a summary of recent programs to improve those key databases. A user's perspective on the quality of the administrative data on CHIP was provided by David Rousseau of the Kaiser Family Foundation. Richard Strauss, of CMS, discussed the quality of the administrative and survey data used for programmatic purposes, including the allocation of funds to states, which was a major use of the data prior to the passage of the CHIP reauthorization in 2009.

In the second panel, Sharon Long of the Urban Institute, John McInerney of the Commonwealth Institute, and Lynn Blewett of the University of Minnesota (also a member of the steering committee) discussed the important state uses of administrative and survey data for state purposes, such as program management and evaluation. The third panel focused on modeling strategies for improving estimates, with presentations by Mark Bauder and Brett O'Hara summarizing the Census Bureau's Small Area Health Insurance Estimation program and William Bell discussing the Census Bureau's Small Area Income and Poverty Estimation model. Nathaniel Schenker of the National Center for Health Statistics discussed joint modeling of survey and administrative record data, using health-related data collections.

On the second day, workshop participants heard from John Czajka of Mathematica Policy Research, Inc., who discussed measuring income; Joanna Turner of the University of Minnesota, who discussed measuring health insurance coverage; and Genevieve Kenney and Victoria Lynch of the Urban Institute, who gave an overview of survey characteristics that would commend or limit their use in estimating coverage of health insurance for children. Finally, we heard from Cindy Mann, director of Medicaid and state operations at CMS, who put the work of the committee into an overall perspective and provided the workshop with a provocative wish list of data needs from the viewpoint of the administrator of CHIP. These presentations were uniformly well prepared, well presented, and provocative.

In the course of the workshop, several important issues were raised on the scope and content of the surveys and administrative databases, which are ancillary to coverage issues. Those topics included the need to collect data on the health and other socioeconomic characteristics of children by insurance status, and the benefits available to children within their insurance coverage (e.g., dental and mental health coverage), special therapies, and the like. In view of the concentrated focus of the workshop on means of estimating coverage, these issues were not explored in any depth. However, these important issues of scope and content should be considered in any future review of these data sources.

The steering committee also acknowledges the excellent work of the staff members of CNSTAT and the NRC for their support in developing and organizing the workshop and preparing this report. This report was prepared under the direction of Constance Citro, director of CNSTAT. Tom Plewes served as study director as well as rapporteur for the workshop. The steering committee was ably assisted in all administrative arrangements by Anthony Mann, also on the CNSTAT staff. Anthony played a major role in preparation of this workshop summary.

This workshop summary was reviewed in draft form by individuals chosen for their diverse perspectives and technical expertise, in accordance with procedures approved by the Report Review Committee of the NRC. The purpose of this independent review is to provide candid and critical comments that assist the institution in making its report as sound as possible, and to ensure that the report meets institutional standards for objectivity, evidence, and responsiveness to the study charge. The review comments and draft manuscript remain confidential to protect the integrity of the deliberative process.

The panel thanks the following individuals for their review of this report: Martha Heberlein, Georgetown Center for Children and Families; Leighton Ku, Center for Health Policy Research, School of Public Health and Health Services, The George Washington University; and Paul W. Newacheck, Department of Pediatrics and Philip R. Lee Institute for Health Policy Studies, University of California at San Francisco.

Although the reviewers listed above have provided many constructive comments and suggestions, they were not asked to endorse the conclusions or recommendations, nor did they see the final draft of the report before its release. The review of this report was overseen by Lisa Lynch, The Heller School for Social Policy and Management, Brandeis University. Appointed by the NRC, she was responsible for making certain that the independent examination of this report was carried out in accordance with institutional procedures and that all review comments were carefully considered. Responsibility for the final content of the report rests entirely with the author and the NRC.

V. Joseph Hotz, *Chair*
Steering Committee for a Workshop on Evaluating Databases for Use in the Children's Health Insurance Program (CHIP) Allocation Formula

Contents

PART I WORKSHOP SUMMARY

PART II BACKGROUND PAPERS

Part I

Workshop Summary

1

Introduction

Each year the federal government distributes more than $12 billion under the Children's Health Insurance Program (CHIP). The program was established by the Balanced Budget Act of 1997 to expand health insurance coverage to uninsured children in families with incomes that are modest but too high to qualify for Medicaid. The program is financed jointly by the federal government and the states, and it is administered by the states within broad federal guidelines. States offer this expanded coverage through Medicaid program coverage expansions or stand-alone programs aimed specifically at providing health coverage to children.

The federal CHIP monies are distributed to the states on the basis of a formula that, until recently, included the number of uninsured low-income children in each state. Through 2008, federal CHIP allocations were calculated using a formula with two key components: the child component factor and the health cost factor. The child component factor was a combination of the number of low-income children (defined as under 200 percent of the federal poverty level, FPL) and the number of low-income uninsured children based on 3 years of pooled state estimates from the Annual Social and Economic Supplement to the Current Population Survey (CPS). The health cost factor, which was used as a proxy for estimated program expenses, was based on estimates by the Bureau of Labor Statistics of the ratio of the average state wage in the health services industry relative to the national wage for the 3 most recent years of available data. States must contribute to the CHIP cost, and the federal government provides matching payments to the states up to their available annual capped federal allotments. The matching rates are based on the Medicaid Federal Medical Assistance Percentages but are "enhanced," reflecting greater federal financial participation.

In early 2009 the Children's Health Insurance Program Reauthorization Act (CHIPRA) changed that allocation practice. For 2009 through 2013, CHIPRA mandates new allotment formulas for each of the 50 states and the District of Columbia as a function of their past allotments, their past or projected federal CHIP spending, or both.

CHIPRA also mandates in Section 602 that the secretary of commerce, in consultation with the secretary of health and human services, develop more accurate state-specific estimates of the number of children enrolled under Medicaid or CHIP, improve the estimates of the child population growth factor, and assess whether the American Community Survey (ACS) produces more reliable estimates for the purposes of the act.

The 2010 health care legislation included a technical correction to Section 2109(b)(2)(B) of the Social Security Act (42 U.S.C. 1397ii(b)(2)(B)), as added by Section 602 of CHIPRA, which strikes "the child population growth factor under section 2104(m)(5)(B)" and inserts "high-performing State under section 2111(b)(3)(B)," which also set a definition of a "high-performing state." A high-performing state is one that "on the basis of the most timely and accurate published estimates of the Bureau of the Census, ranks in the lowest 1/3 of States in terms of the State's percentage of low-income children without health insurance" (Section 2102 (a)(6)). The federal matching payment will be determined in part on whether a state meets this benchmark.

Thus, although CHIP funds are no longer allocated on the basis of coverage estimates, it is presumed that the results of that mandated assessment could be used in assessing the appropriateness of a future CHIP funding formula—one that includes estimates of the low-income uninsured population of children—as well as to meet other specific provisions of CHIPRA and the new health reform legislation that require estimates of health insurance coverage for children primarily to assess state performance and assist in program evaluation.

WORKSHOP FOCUS

To address the provisions of CHIPRA that call for an assessment of the accuracy and adequacy of databases on which estimates of children's health insurance coverage are based, the Office of the Assistant Secretary for Planning and Evaluation of the U.S. Department of Health and Human Services requested that the Committee on National Statistics of the National Research Council convene a steering committee of experts to organize, commission papers for, and conduct a public workshop to critically examine the various databases that could provide national and state-level estimates of low-income uninsured children and could be effectively used as criteria for monitoring children's health insurance coverage. In designing the workshop, the steering committee determined that the data sets for consideration at the workshop would include ACS, CPS, the Small Area Health Insurance Estimates, the National Health Interview Survey (NHIS), and administrative databases.

The workshop was held in Washington, DC, on June 17-18, 2010. The agenda and a list of participants appear in Appendix A. Among the participants were representatives of public policy research organizations and federal agencies with responsibility for CHIP and the databases under consideration.

The exchange of information and the publication of this report with the background papers were the sole goals of the workshop. This report is intended as a record of the discussion of key issues identified by the steering committee and discussed by the subject-matter experts who participated in the workshop. It draws no conclusions, nor does it make any recommendations.

It is important to be specific about the nature of this report, which documents the information presented in the workshop presentations and discussions. It lays out the key ideas that emerged from the workshop and should be viewed as an initial step in examining the research and applying it in specific policy circumstances. The report is confined to the material presented by the workshop speakers and participants. Neither the workshop nor this summary is intended as a comprehensive review of what is known, although it is a general reflection of the field. The presentations and discussions were limited by the time available for the workshop.

This report was prepared by a rapporteur and does not represent findings or recommendations that can be attributed to the steering committee. Indeed, the report summarizes views expressed by workshop participants, and was not designed to generate consensus conclusions or recommendations but focused instead on the identification of ideas, themes, and considerations that contribute to understanding.

REPORT OVERVIEW

The report is organized into two parts. Part I is a summary of the workshop. Following this introductory chapter, Chapter 2 continues with a discussion of the context of the changing environment for the children's health insurance program. Both the 2009 reauthorization of CHIP and the new health reform legislation are bringing significant changes to the program, generating new requirements for good data on coverage. In addition, several initiatives, such as the secretary of health and human services' challenge and program management goals, are also creating demand for good coverage estimates.

Chapter 3 discusses the main federal surveys for measuring health insurance coverage for children: CPS, ACS, and NHIS. Two additional surveys are also discussed: the Survey of Income and Program Participation and the Medical Expenditure Panel Survey, Household Component. Administrative databases are also important elements in the measurement of health insurance coverage for children in that they define the population and its characteristics and reflect enrollments. These state-gathered, federally maintained data collections are discussed in Chapter 4.

With primary responsibility for managing the programs, many states have mounted their own data collections for measuring the health insurance coverage of children in their states. These survey-based collections, some of which are quite extensive, are described in Chapter 5.

Chapter 6 introduces a somewhat different approach to estimating coverage, summarizing three presentations on the main modeling strategies for improving estimates using multiple data inputs. Two Census Bureau models are designed to provide reliable data for poverty and health care coverage: the Small Area Income and Poverty Estimation and the Small Area Health Insurance Estimation projects. The chapter also describes another approach, which involves the combination of data from multiple surveys to develop data that are lacking or of insufficient reliability from any single survey.

Chapter 7 summarizes the rich and stimulating discussion by the participants in response to the question: What do we need to know? In this time of change in health care policy, the way ahead is indeed challenging. The participants' contributions are organized around the broad topics of matching data, gathering data from both public and private sources, rationalizing the ACS and the CPS, and being prepared to assess the impact of the new health care legislation.

Part II consists of six papers that were prepared as background for the workshop. Two appendixes complete the report: Appendix A presents the workshop agenda and list of participants; Appendix B presents biographical sketches of steering committee members.

2

The Changing Policy Context

The need for an accurate accounting of the coverage of children in public and private health insurance programs has been an integral part of the Children's Health Insurance Program (CHIP) since its inception. Since its passage in 1997 as part of the Balanced Budget Act, the goal of the program has been to close coverage gaps facing low-income families who do not have access to affordable private coverage for their children but do have incomes too high to qualify for Medicaid. These legislative and programmatic goals have created a need for good measures of insurance coverage for children.

LEGISLATIVE REQUIREMENTS FOR ESTIMATES OF INSURANCE COVERAGE

CHIP began as a block grant to states (see Kenney and Lynch, Chapter 8 in this volume). With higher federal matching rates than states typically receive under Medicaid, the program gained wide acceptance. Importantly, from the beginning, states have had considerable flexibility to design their CHIP, and as a result eligibility thresholds, outreach efforts, retention, enrollment policies, benefits, and cost-sharing have varied. This flexibility has affected the relative coverage from state to state.

After 1997, CHIP and coverage grew rapidly. However, as Kenney and Lynch point out, despite this progress, at the time when the Children's Health Insurance Program Reauthorization Act (CHIPRA) was passed in 2009, millions of children were still uninsured despite being eligible for Medicaid or CHIP. In an effort to address these gaps, CHIPRA provided states with new tools to address shortfalls in enrollment as well as access and quality. CHIPRA included new outreach and enrollment grants and bonus payments to states that adopted five of eight enrollment/retention strategies and that experienced Medicaid enrollment that exceeded targeted growth rates.[1] States were also given greater options to use "express lane" eligibility strategies to facilitate eligibility determination and enrollment and were given new options for meeting documentation requirements. CHIPRA allowed states to use federal dollars to cover legal immigrant children who had been in the United States less than 5 years (previously coverage for such children had to be funded exclusively with state funds). In her presentation at the workshop, Genevieve Kenney pointed out that, just a year later, data from the Georgetown Center for Children and Families indicate that as many as 15 states actually have expanded eligibility to higher income levels since CHIPRA was passed. Another 19 are now using federal dollars to cover legal immigrant children, and almost 20 have adopted some type of major change in their enrollment and retention processes aimed at increasing participation.

CHIPRA also provided states with additional federal allotments for their CHIP to cover the costs of enrolling more eligible children and of expanding eligibility (e.g., to higher income groups). In addition, CHIPRA included a number of provisions designed to improve access to care and the quality of care for the children served by Medicaid and CHIP.

CHIPRA had two expected outcomes that help define the context for measurement, according to Kenney and Lynch (see Chapter 8). The first is the expected decrease in the number of uninsured. There was indeed an enrollment increase, about a year later, of 2.6 million children (over the period October 1, 2008, to September 30, 2009) (U.S. Department of Health and Human Services, 2010a, p. 1). The second was an expected decline in private coverage, which would partially offset the positive impact on coverage to the extent that CHIP gains were accompanied by cuts in private insurance coverage.

In their paper, Kenney and Lynch suggest that an independent source

[1] The strategies are (1) adopting 12-month continuous eligibility for all children, (2) eliminating the asset test for children, (3) eliminating in-person interview requirements at application and renewal, (4) using joint applications and supplemental forms and the same application and renewal verification process for the two programs, (5) allowing for administrative or paperless verification at renewal through the use of prepopulated forms or ex parte determinations, (6) exercising the option to use presumptive eligibility when evaluating children's eligibility for coverage, (7) exercising the new option in the law to use the Express Lane Eligibility option, and (8) exercising the new options in the law in regard to premium assistance (Georgetown Center for Children and Families, 2009).

of data on coverage is needed to answer the following evaluative questions, important for gaining an understanding of the impact of CHIPRA:

- Did uninsured rates fall among children following enactment of CHIPRA? If so, by how much?
- How much do Medicaid/CHIP participation rates vary across states? Do differences across states narrow over time?
- How did rates of public and private coverage change for different groups of children (defined by race/ethnicity, income, age, health status, etc.)?
- To what extent are the observed changes in uninsurance, public coverage, and private coverage among children attributable to CHIPRA?

With much the same emphasis, Chris Peterson pointed out that Congress is broadly interested in good data because of an interest in measuring both need and success. Although need is expressed in terms of funding, it also drives decisions about other resources, for example, whether or not to build hospitals, clinics, or other infrastructure. The questions of need add a geographic dimension to the data because many of the resource decisions are made at the local level. The use of data in the measurement of success is also straightforward. Congress wants to know if the programs that have been implemented have been successful. For CHIPRA, has it resulted in a reduction in uninsurance among children?

The ink was barely dry on the CHIPRA legislation when it was modified by two other important pieces of legislation. The American Recovery and Reinvestment Act (ARRA) includes enhanced matching rates to states that maintain their Medicaid eligibility thresholds for children and adults, in an effort to induce them to continue supporting Medicaid coverage during the current recession. The enhanced matching rates, which are at least 6.2 percentage points higher than regular matching rates, were implemented on October 1, 2008; they are slated to continue through June 2011, with a phased-down increase in the matching support.

The Patient Protection and Affordable Care Act of 2010 (PPACA, P.L. 111-148) contains a number of important policy changes that could affect both Medicaid and CHIP coverage for children. It legislates comprehensive health reform, including an expansion of Medicaid to adults and children up to 133 percent of the federal poverty level (FPL) by January 2014, a maintenance of effort requirement through 2019 on state Medicaid and CHIP coverage for children, the provision of new subsidies for the coverage of families with incomes up to 400 percent of the FPL, the creation of health insurance exchanges, and coverage mandates for both individuals and employers. (See Box 2-1 for a summary of coverage provisions of

BOX 2-1
Coverage for Children Under CHIPRA and PPACA

Infants under age 1 and children aged 1-5*
- Medicaid < 133% of the federal poverty level
- CHIP 133-200% of the federal poverty level

Children aged 6-18*
- Medicaid < 100% of the federal poverty level, going to 133% in 2014
- CHIP 100-200% of the federal poverty level

Disabled children: Medicaid medically needy option

Other children: Foster care, children aged 18-20

*States can disregard income and extend coverage to higher percentages of the federal poverty level (e.g., 300% of the federal poverty level)
SOURCE: Baugh (2010).

CHIPRA and PPACA.) PPACA also provided 2 additional years of federal funding for CHIP, beyond what was in CHIPRA, through 2015.

As noted earlier, PPACA includes a technical correction that also refers to definition of a "high-performing state"—one that "on the basis of the most timely and accurate published estimates of the Bureau of the Census, ranks in the lowest 1/3 of States in terms of the State's percentage of low-income children without health insurance." The federal matching payment is determined in part on whether a state meets this benchmark.

EVOLVING CRITERIA FOR STATE FUNDING ALLOCATIONS

Throughout its history, CHIP has relied in some way on estimates of children's insurance coverage, mainly in regard to establishing the amounts allotted to the states for program operations and evaluation. The criteria for allocating funds to the states have changed over the course of the program, and so has the need for coverage estimates.

In his presentation to the workshop, Richard Strauss outlined the past and current practices used by the Centers for Medicare & Medicaid Services (CMS) for allocating funds to states for CHIP. Prior to 2008, the fiscal year allotment (A) for the states (including the District of Columbia) was based on the number of children (Ci) under age 19 with a family income equal to or less than 200 percent of the FPL and the number of such children with no health insurance. A state cost factor (SCFi) for each state was annual wages in the health industry for the state (Wi) and annual wages in the health industry for all states (WN). The allotment formula for fiscal years 1998 through 2008 was given as:

$$A \times [\ (Ci \times SCFi)\ /\ \Sigma(Ci \times SCFi)\ ].$$

Strauss pointed out that the statute was very specific as to the source of the data that determined the number of children in the state. The number was determined on the basis of data as reported and defined in the three most recent March supplements to the Current Population Survey (CPS) before the beginning of the calendar year in which the fiscal year begins.

The formula allocation procedures that governed the allocation of funds for the first decade of the program were changed with CHIPRA in 2009. CHIPRA based the amount each state would receive for 2009 and subsequent fiscal years on the 2008 allocation as adjusted by an allotment increase factor. This factor is based on a *per capita health care growth factor* (based on the percentage increase in the population of children in the state from July 1 of previous fiscal year to July 1 of the fiscal year involved, as provided by CMS), and a *child population growth factor* (based on the percentage increase in the per capita amount of national health expenditures from the calendar year in which the previous fiscal year ends to the calendar year in which the current fiscal year ends, as provided by the Census Bureau).

Beginning in 2011, the amount will be adjusted by "rebasing." Under rebasing, the allotment for a fiscal year is determined by applying the allotment increase factor to the previous fiscal year's expenditures, which were applied against the allotments that were available in the previous fiscal year. Under CHIP, this alternates with each fiscal year; that is, the allotment increase factor is applied against the previous fiscal year's allotment or the previous fiscal year's expenditures.

COVERAGE ESTIMATES FOR PROGRAM PURPOSES

Concerned over the estimate by the Urban Institute that some 5 million eligible children remained unenrolled a year after enactment of the CHIPRA legislation (based on data from CPS and the CHIP State Enrollment Data System), Secretary of Health and Human Services Kathleen Sebelius issued a challenge to the states called Connecting Kids to Coverage. Her challenge to states, local governments, community-based organizations, health centers, and faith-based organizations was to enroll the estimated 5 million children who are eligible for CHIP or Medicaid but do not have coverage over the next 5 years (Sebelius, 2010).

As part of this initiative, the U.S. Department of Health and Human Services has a plan to engage a contractor to assist with measuring progress in meeting the secretary's challenge. A request for proposals was issued shortly after the completion of the workshop. The department is asking a contractor to analyze survey and administrative data to develop quarterly national estimates of the number of uninsured but eligible children (aged 0-18) and comparable estimates for all states and the District of Columbia. The contractor's report is expected to outline the methodology for the estimates, ensure that the projections will be replicable, and explain the methods clearly in a separate report on methodology (U.S. Department of Health and Human Services, 2010b).

In her workshop presentation, Cindy Mann, director of the Center for Medicaid and State Operations, stated that the secretary's challenge is a high priority for the agency and the states. In order to determine the progress toward the coverage goal for the challenge, the department needs good estimates of insurance coverage by state. Progress will be measured by calling for semiannual reports by state that include current estimates of the number of uncovered children. The measurement of progress toward accomplishing the challenge goal requires current and timely data on the uncovered population by state.

Another issue for which good data is needed is the targeting of outreach grant funds. Mann stated that CHIPRA authorized $100 million in outreach grants, of which $40 million has been granted to community-based organizations, $10 million to states and counties, and $10 million to Indian health organizations. The remaining $40 million and an additional $40 million under PPACA will be targeted by group and geographic area based on evidence. Similarly, CHIPRA provides matching funds for translation and interpreter services, another focus of data to help target intervention strategies.

Mann also identified a wish list for data to assist in managing and evaluating CHIP. The list includes

- Data on the longitudinal experience of low-income children by urban and rural location, composition of their households, immigration status, and language issues. For various reasons, children can enter and leave coverage over the course of a year, and it is important to identify the causes and effects of this kind of transition.
- More timely data, both from the administrative databases and from surveys. The eventual goal would be real-time data on enrollment and uninsurance at the state level.

- Data to better understand gaps in insurance coverage. Mann gave the example of a drop in enrollment in Arizona due to a program freeze in that state. It is important to know how many children are losing coverage because they are unable to renew it, or because they are moving into Medicaid. (Medicaid enrollment is increasing in Arizona.)
- Better data on the status of children beyond enrollment in CHIP. This includes the kind of care they are receiving by income and the impact of that care as measured by diagnosis. These detailed data would answer questions about different patterns of care around the country and different levels of access to certain providers and to certain protocols.
- More standardization of state administrative data and better integration of federal and state data collection systems. In many cases, states are already collecting data useful for program administration, such as on determinations of applications, denials, and determination of renewals, but the data are not required at the federal level, so they are not available to the federal government for program management purposes.

3

Federal Surveys

For the most part, estimates of children's health insurance coverage are based on one of three federal household surveys, although several other federal surveys collect data periodically that assist in understanding the extent and quality of health insurance coverage. The three major surveys are (1) the American Community Survey (ACS), (2) the Current Population Survey Annual Social and Economic Supplement (CPS ASEC), and (3) the National Health Interview Survey (NHIS). (The attributes of these three major surveys are summarized in Table 3-1, and are discussed below.) Two other surveys are referenced here because they, too, collect insurance coverage information: the Survey of Income and Program Participation (SIPP) and the Medical Expenditure Panel Survey-Household Component (MEPS-HC).

NATIONAL HEALTH INTERVIEW SURVEY

Kenney and Lynch report a general consensus that the NHIS produces the most valid coverage estimates (Chapter 8, in this volume).[1] The NHIS is a multipurpose health survey conducted by the National Center for Health Statistics (NCHS), Centers for Disease Control and Prevention (CDC), and is the principal source of information on the health of the civil-

---

[1] Results from the Medicaid Undercount project suggest that underreporting of Medicaid/CHIP is lower in the NHIS than the CPS (see http://www.census.gov/did/www/snacc/ [September 2010]).

TABLE 3-1 Comparison of the Attributes of Three Sources of Coverage Estimates

Strengths	Limitations
American Community Survey	
• Very large sample • Point-in-time coverage variable • Geographic identifiers • Seems to support fairly robust eligibility simulations	• Mostly mail mode • Just an itemized list asking about coverage • No attempt to correct for misreported coverage (in 2008) • Coverage question is new (2008) so preperiod analysis is limited • Difficult to customize questions or wording
Current Population Survey	
• High-quality data collection methods • Detailed questions about coverage • Attempts to correct for misreported coverage • Supports robust eligibility simulations • State-level identifiers • Long-time series allows expansive preperiod analysis • Questions can be customized	• Previous year reference period for coverage question increases recall bias and confusion • Sample size and number of primary sampling units small in some states
National Health Interview Survey	
• High-quality data collection methods • Detailed questions about coverage • Attempts to correct for misreported coverage • Time series available • State-level identifiers available	• Sample size and number of primary sampling units too sparse • No state identifiers in public-use microdata file • Quality of income data not certain
Combined	
• Comparisons help reveal measurement errors • Broad range of coverage measures and related variables available for study	• Published estimates difficult to compare • Not enough explanation of why estimates differ

SOURCE: Kenney and Lynch (Chapter 8, in this volume) and workshop discussion.

ian, noninstitutionalized, household population of the United States. The NHIS has been conducted continuously since 1957; the data are collected for NCHS by the Census Bureau (National Center for Health Statistics, 2010). The NHIS is collected in four segments that are weighted, separately, to February 1, May 1, August 1, and November 1 of the survey year, then combined to create a single annual weight in the public-use file, with an effective reference date of mid-June (see Chapter 10, in this volume).

Because the NHIS is a health-focused survey, it includes many features to aid respondents in understanding the coverage question and recalling details required to correctly answer it. Kenney and Lynch consider the NHIS features that may strengthen validity to include

- the area sample frame;
- a well-trained interview staff that work exclusively on this survey;
- a fairly high response rate;
- usually an in-person interview;
- a knowledgeable respondent (i.e., adults are encouraged to report on coverage for children if they are familiar or to talk about it with the person most familiar with the coverage of household members);
- a questionnaire that defines concepts and probes respondent memory as it collects information;
- breadth of content on other health-related data, which potentially helps respondents understand distinctions between coverage types and their type;
- asking about coverage source at the time of the survey, which is associated with lower measurement error;
- asking about Medicaid and CHIP using state-specific names;
- a low level of item nonresponse on insurance sequence;
- asking for many details about coverage, which may help define relevant concepts or distinguish different types of coverage (e.g., type of managed care, copayments, deductibles, need for referrals) and help respondent recall coverage details;
- asking about periods without coverage and when the child last had coverage (for use in estimating full-year uninsurance) and why it stopped (potentially helping the person to recall more details required to determine their true coverage status);
- verifying no Medicaid for children with no reported coverage;
- asking about citizenship, place of birth, and family relationship, which are some of the important variables needed to simulate eligibility in Medicaid and CHIP;
- asking about medical visits and other uses of coverage or evidence of acting uninsured;
- asking for the name of the insurance plan so the name can be matched to a list of insurance plans by state in a postcollection data processing phase and used to recode misreported coverage type; and

- having been in production for many years and with attention to maintaining a credible time series.

The authors caution that there are also important limitations for using the NHIS to monitor coverage. The most problematic of these is the sampling design, which limits the geographic and other subpopulation estimates that are possible as well as raising validity questions. First, the sample size is too small to produce precise annual state (and substate) estimates for most states. Second, most states have only a very small number of primary sampling units, a fact that raises a concern about the representativeness and precision of the state-level estimates produced by the survey. Third, because of data confidentiality concerns, access to state identifiers is available only through data centers.

The ability to use the NHIS to simulate Medicaid/CHIP eligibility is also limited by the quality of the income data, as well as the possible underreporting of the Medicaid/CHIP information coverage, despite all the efforts to measure coverage accurately. The NHIS is also limiting because of the timing of the data release and what is excluded from the published estimates. There is an early release that enables some important coverage evaluations before the survey is fully prepared; however, it is still about 9 months after the interviews are completed, it does not include published estimates for children ages 0-18 separately, and it does not provide valid estimates by income.

CURRENT POPULATION SURVEY

The CPS ASEC has historically played an important role in monitoring coverage. Besides being a relatively large survey that uses high-quality data collection methods, it is an income- and employment-focused survey and is considered by the Census Bureau to have the most valid data on those domains, which are integral for eligibility simulations and other coverage-related analyses. It is the source of the national poverty estimates.

Kenney and Lynch suggest that features that may strengthen validity include

- the area frame;
- a well-trained interview staff working exclusively on this survey;
- telephone and in-person interviews;
- high response rates;
- a sample size that is large enough for precise state estimates for large states annually;
- family-level questioning about coverage source, which helps get more coverage reported in large households;

- state-specific names and separate questions about Medicaid and CHIP;
- other probes and definitions (military coverage, direct purchase);
- a question on directly purchased coverage that emphasizes that it is not related to a current or former employer;
- asking for detailed information about coverage, including who is the policy holder, who is covered by the same policy, who is covered by someone outside the household, and employer contributions;
- asking several times about any other type of coverage not yet talked about;
- verifying the absence of insurance coverage;
- logical coverage edits performed by the Census Bureau to correct some likely reporting errors;
- asking about citizenship, place of birth, family relationship, supports from people outside the household, firm size, as well as income and employment-related factors, which are some of the important variables needed to simulate eligibility in Medicaid and CHIP;
- asking about health status;
- having been in production for many years and with attention to maintaining a credible time series; and
- the release of estimates and public-use files with state identifiers 5-6 months after the data are collected.

The CPS ASEC has limitations as well, the most critical of which for purposes of monitoring coverage is the known measurement error with the coverage questions because of confusion, recall bias, and other issues with the retrospective reference period (Pascale et al., 2009).

The list of limitations suggested by Kenney and Lynch goes on: it has a small sample size for many individual states, even when averaged over 3 years, which makes state estimates highly variable; the imputations for missing income and health insurance data (fully 30 percent of the CPS income data is imputed) are developed on a national basis and do not take account of state differences; income is underreported, particularly for the low-income population, which may contribute to overestimates of children eligible for CHIP; there are errors in recalling periods of health insurance coverage, so that estimates of children with no coverage at any time during the previous year may be biased upward; and children, particularly girls and minorities, are undercovered.

The extent to which nonsampling errors in the CPS were a problem for fund allocations, such as underreporting of income and health insurance and undercoverage of the population, depended on the extent to which the various sources of error differed among states and thereby under- or overcompensated particular states in terms of their CHIP allotments relative to other states.

As a result of these limitations, the CPS has greater apparent underreporting of Medicaid/CHIP coverage and considerable uncertainty about what the estimates mean, compared with other surveys (Davern et al., 2009). In addition, the sample size and the number of primary sampling units is small in many states, which raises concerns about the representativeness and precision of the state estimates.

Historically there has also been concern about bias in the imputation process for coverage variables; however, new imputations are being implemented at the Census Bureau to address this problem (Davern et al., 2007). The CPS is also missing information about access to, need for, and use of health services and spending. The 9-month lag between the end of the calendar year reference period and the release of the published estimates and edited data limits the ability to track coverage in real time; however, the estimates and data are more timely if they are interpreted as representing some time closer to the interview data in March (just 6 months before the release). The published estimates are also limiting because they do not include children ages 0-18.

AMERICAN COMMUNITY SURVEY

ACS, an annual survey designed to provide intercensal estimates of the information previously collected on the decennial census long form, added information on health insurance coverage in 2008. Although the ACS is a new resource for monitoring children's health insurance coverage, it has a number of important strengths relative to the other surveys, the most important of which is its very large sample and its sample frame (which samples every county and census tract). The large sample size allows for 1-year coverage estimates for areas with a population of 65,000 or more; 3-year coverage estimates for areas with populations of 20,000 or more; and 5-year coverage estimates for all statistical, legal, and administrative entities. Thus, it is possible to put together a variety of substate estimates, including service areas consisting of public-use microdata areas, counties, metropolitan areas, and the like. It is quite timely, in that estimates and public-use files with state identifiers are released 8-9 months after the end of the survey period, implying an average lag between data collection and data release of 14-15 months. In addition, although most data are collected by mail, the ACS has very high response rates (98 percent).

The health insurance coverage question was asked for the first time in 2008 and is asked for each person in the household. The respondent is instructed to report each person's current coverage by marking "yes" or "no" for each of the eight types of coverage listed in the question:

> Is this person CURRENTLY covered by any of the following types of health insurance or health coverage plans? Mark "Yes" or "No" for EACH type of coverage in items a–h.
>
> a. Insurance through a current or former employer or union (of this person or another family member)
> b. Insurance purchased directly from an insurance company (by this person or another family member)
> c. Medicare, for people 65 and older, or people with certain disabilities
> d. Medicaid, Medical Assistance, or any kind of government-assistance plan for those with low incomes or a disability
> e. TRICARE or other military health care
> f. VA (including those who have ever used or enrolled for VA health care)
> g. Indian Health Service
> h. Any other type of health insurance or health coverage plan—Specify [Includes a space to write-in a response to subpart h]

Since the ACS coverage data are new as of 2008, the survey cannot provide an extended preperiod for studying trends in children's coverage prior to the adoption of policy changes related to the Children's Health Insurance Program Reauthorization Act (CHIPRA). In addition, reliable data will not be available immediately for smaller local areas because, as noted above, the Census Bureau publishes ACS single-year estimates for areas with a population of 65,000 or more, 3-year estimates for areas with populations of 20,000 or more, and 5-year estimates will be published for all statistical, legal, and administrative entities. This means that the first 3-year estimates will be released in 2011, and the first 5-year estimates will be released in 2013.

Moreover, research is just now being conducted on the validity of the ACS estimates. Although overall the unadjusted ACS estimates of uninsured children were close to the CPS estimates, they were somewhat higher than the NHIS estimates for the same period, and reports of direct purchase of insurance on the ACS are very high. A major limitation of the survey is that it is conducted primarily by mail (56 percent), which means that most respondents complete the survey without the aid of an interviewer.

Another major concern is that the coverage question is limited. It includes no distinction among Medicaid, CHIP, and other sources of government insurance. Several times during the workshop, the ACS was criticized for not permitting the inclusion of state-specific names for Medicaid or CHIP. In response to a suggestion that they be added, David Johnson told workshop participants that earmarking questions on the ACS to a particular state is nearly impossible for the Census Bureau because of the cost of sending out tailored surveys by state. However, this information was made available to interviewers in the computer-assisted telephone interviewing (CATI) and computer-assisted personal interviewing (CAPI) modes starting in 2009. There was some discussion about the cost-effectiveness of tailoring the questionnaire and some interest in studying the costs and benefits of such an enhancement.

Another ACS limitation is that there is only one itemized list of coverage types (rather than a detailed series of patterned questioning, defining, and probing, as in the NHIS and the CPS), which could also introduce more measurement error in the reporting of coverage type. Also of concern is the absence of a statement that insurance purchased directly should not have anything to do with a current or former employer as well as the absence of questions about coverage details (managed care, premiums, employer contributions). In addition, the ACS does not include a verification of uninsurance or questions about duration of uninsurance.

Another concern with the 2008 ACS estimates is that there was less postcollection processing on the ACS to remedy possible reporting errors, compared with the NHIS (which collects the name of the person's plan and then uses it to reclassify coverage type) and the CPS (which uses other coverage-related information collected about the person or family to reclassify coverage on a logical basis).

In her presentation, Joanna Turner suggested that this latter concern shows up in the treatment of nonresponse imputation and editing of the new coverage question. Although 73 percent of respondents had a complete item response (that is, a "yes" or "no" to each of the first seven types), 23.2 percent had a partial response (responded to at least one, but not all items), and 3.8 percent left all items blank. Response patterns varied among the collection modes. Mail respondents were the least likely to provide complete item response, at 58.1 percent, and those interviewed through CATI and CAPI were the most likely to give complete item response, at 96.1 percent. This pattern reflects both differences in the instruments and differences in the composition of people in each mode.

When respondents do not complete an item, a series of rules determine a response that is imputed to the record. If a respondent marked "yes" to one and only one of the types and all other subparts were left blank, the types associated with the blanks were assumed to be and assigned values of "no." For example, if a respondent marked "yes" for employer-provided coverage (subpart a) and left the rest blank, the edited final response for that person would be a "yes" for employer- or union-based coverage and a "no" for all of the others: direct purchase,

Medicare, Medicaid, military health care, veterans, and Indian Health Service. This process turned some partial responses into complete responses, and they were not considered imputed. This editing choice was the result of analysis of respondents to the paper form. The weighted allocation rate—the percentage of people who had an answer to at least one of the health insurance types, obtained through a procedure called hot-deck imputation—was 9.7 percent for 2008. (The ACS edits for nonresponse did not use a rules-based assignment of health insurance coverage, called coverage or consistency edits. In the ACS, these types of edits are being implemented in the 2009 tabulations.)

The ACS has a number of other content limitations. In particular, family relationship information is not directly available for analysis, making it much more difficult to identify health insurance units for eligibility simulations; there is no information on the child's general health status or the parents' firm size. And although the ACS sample is very large and its published estimates cover a variety of important geographic areas (e.g., congressional districts), the sample released for public-use data is smaller and excludes many geographic identifiers, making it more difficult to track meaningful coverage changes for smaller states and smaller subgroups, short of gaining access to a census data center (which requires a comprehensive application that takes several months and must meet stringent requirements).

MEDICAL EXPENDITURE PANEL SURVEY, HOUSEHOLD COMPONENT

The Medical Expenditure Panel Survey (MEPS) has two major components: the Household Component (MEPS-HC), which provides data from individual households and their members supplemented by data from their medical providers, and the Insurance Component, which is a separate survey of employers that provides data on employer-based health insurance. MEPS-HC collects data from a sample of families and individuals in selected communities across the United States, drawn from a nationally representative subsample of households that participated in the prior year's NHIS. During the household interviews, MEPS collects detailed information for each person in the household on the following: demographic characteristics, health conditions, health status, use of medical services, charges and source of payments, access to care, satisfaction with care, health insurance coverage, income, and employment. It is a panel design, that is, there are several rounds of interviewing covering 2 full calendar years, making it possible to determine how changes in respondents' health status, income, employment, eligibility for public and private insurance coverage, use of services, and payment for care are related (Agency for Healthcare Research and Quality, 2010).

The health insurance section of the MEPS collects extensive information about private health insurance obtained through an employer, direct purchase of private insurance plans, and public health insurance programs, including on Medicare, Medicaid/CHIP, Medicaid waiver programs, TRICARE, Veterans Administration coverage, and other government programs. It identifies the household members covered by health insurance, type of plan, name of each plan, nature of coverage under each plan, duration of coverage, and who pays various costs for the policy premiums.

The MEPS has some technical anomalies that should be taken into account, according to Thomas Selden. In response to a question, he suggested that users use the full-year MEPS file to analyze coverage because the full year aligns on poverty and the point-in-time file does not. The point-in-time estimates from the MEPS are very high partly because the Agency for Healthcare Research and Quality changed the sample frame along with the NHIS and revised the computer software used to administer the computer-assisted interviewing. Their research has identified, not a major reason for the difference, but rather many small influences.

SURVEY OF INCOME AND PROGRAM PARTICIPATION

The SIPP is currently designed to collect source and amount of income, labor force information, program participation and eligibility data, and general demographic characteristics to measure the effectiveness of existing federal, state, and local programs, among other purposes. It is a household panel, that is, the survey design is a continuous series of national panels, with sample size ranging from approximately 14,000 to 36,700 interviewed households, each of which ranges in duration from 2.5 to 4 years. The SIPP sample is a multistage stratified sample of the U.S. civilian noninstitutionalized population. Beginning with the 2004 version, the SIPP collects point-in-time ("last 4 months") information on participation in CHIP, although it does not reference state CHIP by name, so there may be some confusion about the program. The most recent data from the SIPP is for the year 2008. Johnson reported at the workshop that the program is undergoing a redesign, in which the Census Bureau is testing moving from collection three times a year to once a year, in the process asking the respondents to complete a calendar indicating when they have received the benefits. The tests, he reported, are promising in terms of the rates reported.

SURVEY COVERAGE ESTIMATES

Turner stated that, in terms of national estimates of coverage rates, the results from the three major sources are very similar. The ACS health insurance coverage rate was 84.9 percent, not statistically different from the NHIS rate of 85.2 percent. This high level of consistency is a good sign for the ACS, which is conceptually similar to the NHIS, as they both measure current coverage. The CPS ASEC health insurance coverage rate was 84.6 percent. Although the statistical test of the difference between the ACS and the CPS ASEC showed evidence of difference, these two estimates do not appear meaningfully different—both round to 85 percent of the population. Both the ACS and the CPS ASEC estimate 90.1 percent of children have health insurance coverage, and the NHIS estimates that 91.0 percent of children have health insurance coverage.

Looked at another way, however, the three surveys show quite different results on how many children are uninsured at a point in time or throughout the year. Kenney and Lynch reported that, for 2008, the most recent year for which official estimates are available from each of these surveys, the number of uninsured children aged 0-17 at a particular point in time ranges from 6.6 million on the NHIS to at least 10.7 million on the MEPS (Turner et al., 2009). The unadjusted CPS and ACS published estimates are both 7.3 million for children aged 0-17 in 2008, whereas they are 8.2 million and 8.1 million, respectively, for the more policy-relevant 0-18 population (Turner et al., 2009).

The most recent year for which full-year estimates of uninsurance are available for more than one survey (assuming the CPS estimate is not a valid measure of full-year uninsurance) is 2007, and they show a range from 3.7 million in the NHIS to 7.9 million in the MEPS (Agency for Healthcare Research and Quality, 2009; Turner et al., 2009). Not only is there disagreement about how many children lack health insurance coverage at a particular point in time nationally, but state-level estimates also vary across surveys (Blewett and Davern, 2006; Turner et al., 2009).

In his presentation, Michael Davern put the estimates into further perspective, drawing on the data on uninsurance for the total population, which is free from some of the issues affecting the estimates of children. Estimates of the percentage and number of people under age 65 who were uninsured throughout 2002 ranged from 17.2 percent (43.3 million) in the CPS to 8.1 percent (20.4 million) in the SIPP. These differences raise some interesting analytical questions:

- Is the range in all-year uninsured coverage estimates common in other domains (poverty, employment, education)?
- Why is all-year uninsured an outlier?
- Is the difference among the surveys' rates in all-year uninsured consistent across important covariates of health insurance coverage?
- Is the CPS consistently like a "point-in-time estimate"?

To address these issues, he examined health insurance coverage estimates as well as other domains that the four national surveys have in common, to see if a common pattern comes through cross-tabulations. The estimates were, as much as possible, anchored to calendar year 2002, but different survey timing caused him to use the 2003 CPS, the 2001 SIPP panel, the 2002 NHIS, and the 2001 MEPS panel. The results of the comparisons for the age group 0-65 years are shown on Table 3-2, which was contained in Davern's PowerPoint presentation.

From this table, Davern observed that the all-year uninsured estimates are an outlier among the other domains examined. The three surveys with an explicit point-in-time estimate are much closer, well within the range of usual survey to survey variation. Furthermore, the CPS is consistently the outlier, with the highest rates of all-year uninsured across the domains examined. His findings suggest that the degree of difference can become dramatic for important policy-relevant groups. For example, the CPS estimate is that 33.6 percent of those below poverty lack insurance, whereas for the SIPP it is 18 percent. As Davern pointed out, this difference is dramatic. This range of 15.7 percentage points is considerably higher than the 9.1 percent difference in the overall estimate. For those whose incomes are over 400 percent of the poverty level, the difference between the SIPP and the CPS narrows to only 5.6 percent points. The difference between the CPS estimate and other all-year uninsured estimates is seen to vary considerably across important domains of poverty, age, race, and education as well. He observed that the CPS is not a reliable measure of all-year uninsured, mainly because it does not perform consistently. He ascribes this partly to the fact that asking people to remember what insurance coverage they had 16 months ago is a difficult cognitive task, especially for programs like Medicaid or CHIP, in which they may be more likely to move on and off the program than in other types of insurance coverage.

Going on to compare the CPS estimate with the NHIS point-in-time estimate across domains, Davern sees significant differences in 27 of 38 domains, with the NHIS showing a consistently lower point-in-time estimate except in two domains—other race and Supplemental Security Income receipt. The MEPS point-in-time estimate across domains shows significant differences in 13 of 38 domains compared with the CPS estimate, indicating that the MEPS moves in a fairly similar fashion to the CPS. Finally, the SIPP point-in-time estimate across domains shows significant differences in 26 of 36 domains, and with the SIPP there is not a very a strong degree of consistency across domains. (Sometimes the SIPP is higher—for children under age 18—and sometimes lower—for adults aged 19-64.)

Davern suggested that, for policy reasons, the fact that persons do not have to be uninsured for 16 months in order to be eligible for state

TABLE 3-2 Comparisons for the Age Group 0-65 Years

Variable	CPS Estimate (%)	NHIS Estimate (%)	Difference (%)
All year uninsured	17.2	9.9	7.3
Uninsured point in time	17.2	15.6	1.6
Special characteristics			
Not born in the U.S.	12.6	11.7	1.0
Born in the U.S.	87.4	88.3	–1.0
Poor health	7.9	6.9	1.0
At least good health	92.1	93.1	–1.0
Student 18-23-years-old	4.5	1.9	2.6
No high school diploma	12.9	13.3	–0.5
High school	29.0	27.5	1.5
Some college	29.5	31.3	–1.8
College graduate	18.7	18.1	0.7
Post-bachelors	9.8	9.7	0.1
Not employed	23.4	20.1	3.3
Employed part-time	9.2	12.7	–3.4
Employed full-time	67.4	67.2	0.1

NOTE: n/a = not available.
SOURCE: Data from Davern (2010).

CHIP or Medicaid means that point-in-time estimates are more reliable in helping to understand who is actually uninsured and who is potentially eligible for a program today, rather than who is among the long-term uninsured. In response to a question, Johnson reported that the Census Bureau has tested a point-in-time question on the CPS and is evaluating the response, with the thought of replacing the all-year estimate with a point-in-time one.

This analysis underscores Davern's observation that the CPS is a very poor measure of all-year uninsurance. Interpreting it as something it was not designed to measure should be done cautiously. There may be a number of causal factors leading the CPS to mimic a point-in-time measure that may not always track an explicit point-in-time measure in a time series. Known causes include editing, imputation, and recall errors. His bottom line is that using the CPS as a point-in-time measure is reasonable (but cautioned), but using it as an all-year measure is not (see Figure 3-1).

MEPS		SIPP	
Estimate (%)	Difference (%)	Estimate (%)	Difference (%)
12.9	4.3	8.1	9.1
17.9	–0.7	15.9	1.3
12.2	0.5	n/a	n/a
87.8	–0.5	n/a	n/a
8.4	–0.5	8.6	–0.7
91.6	0.5	94.1	0.7
4.5	0.0	4.9	–0.4
17.0	–4.1	12.9	–0.1
31.3	–2.3	28.4	0.6
24.1	5.5	31.7	–2.2
16.8	2.0	17.6	1.1
10.9	–1.1	9.3	0.5
20.1	3.3	19.3	4.1
12.2	–2.9	10.6	–1.4
67.7	–0.4	70.0	–2.7

QUALITY OF THE INCOME DATA FROM THE MAJOR FEDERAL SURVEYS

The quality of estimates of health insurance coverage of children, particularly for purposes of CHIP, relies on the quality of the income data collected from the household, which plays a key role in the development of estimates of poverty and low-income status. The paper by John Czajka (Chapter 10, in this volume) suggests that, although much attention is being focused on the quality of these new estimates of health insurance coverage as a major factor in the choice between the CPS or the ACS, the quality of the income data collected in the ACS may be equally important in determining the ultimate viability of the ACS as a source of annual estimates of low-income children and low-income uninsured children.

The CPS, Czajka pointed out, is the official source of annual estimates of income and poverty in the United States, whereas the ACS collects more limited information on personal income and does so through a questionnaire that two-thirds of the sample completes without the assistance of an interviewer and submits by mail. Policy analysts also have reason to be interested in the ACS as a potential data source for a wide variety of analyses of state and local variation in health insurance coverage, and income is likely to play a key role in many such analyses. For such appli-

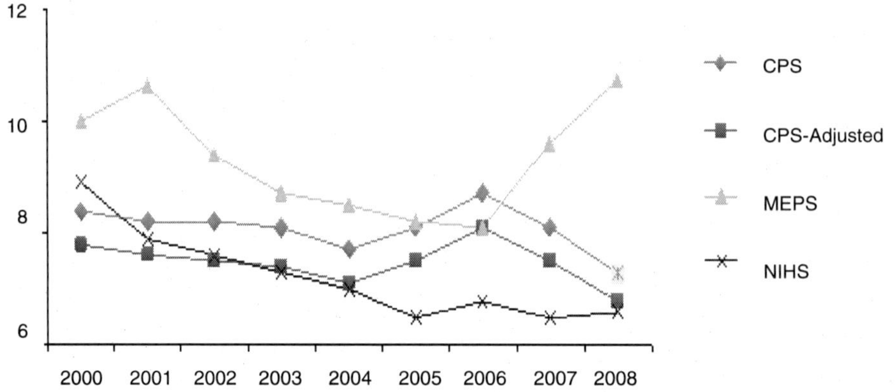

FIGURE 3-1 Trends among the surveys in the number of children (under 18 years) who are uninsured for entire year (CPS) and point-in-time (MEPS and NHIS) (in millions).
SOURCES: CPS–Current Population Survey, 2001-2009; CPS-Adjusted–Davern (2010); MEPS–MEPS-HC online tables, Table 5 (multiple years); NHIS–Cohen et al. (2007) .

cations, the demands placed on the ACS income data are likely to exceed those that must be met in providing satisfactory estimates of low-income children.

Given these potential uses of the ACS and the current limitations, do the ACS income data measure up to the CPS sufficiently well to warrant their use in producing the mandated annual estimates of low-income children and low-income uninsured children? His paper reviews income measurement and other features that differentiate the CPS, the ACS, and the NHIS, presenting findings from two empirical analyses comparing the surveys.[2]

Czajka contends that the CPS defines the measurement of annual income and poverty estimates for the United States since it is the official source of such data. Income in the CPS is defined as pretax money income as measured in the survey; the CPS family is two or more persons living

[2] The first set of findings is drawn from a recent, comprehensive, and systematic assessment of the income data in eight major surveys and their utility for policy-related analyses. This research was conducted by Mathematica Policy Research, Inc., and its subcontractor, Denmead Services and Consulting, under a contract with the Office of the Assistant Secretary for Planning and Evaluation in the Department of Health and Human Services. The second set of findings, which are restricted to the CPS and the ACS, was prepared to examine specific issues related to the use of income data in the analysis of health insurance coverage.

in the same household and related by blood, marriage, or adoption; and the CPS defines residence rules, which affect who is included in a given family. The CPS collects data on the presence of more than 50 sources of income and captures up to 24 annual dollar amounts for each sample member aged 15 and older. The reported incomes of individual family members at the time of the interview are summed to obtain a measure of total family income for the preceding calendar year. By contrast, the ACS collects income for up to eight sources for each sample person aged 15 and older, combining many sources for which the CPS collects separate reports. The NHIS collects total family income from a single question asked of the family respondent. Earnings from employment are collected from all persons aged 18 and older, but this is separate from and not reconciled with reported family income.

There are other differences. The CPS and the NHIS responses are based on income for the previous calendar year, whereas the ACS asks about income for the past 12 months. The ACS income data during a given calendar year span a 23-month period centered on December of the prior year because, for persons completing the survey at the beginning of the year, the past 12 months are January through December of the prior year; for persons completing the survey at the end of the year, the past 12 months are December of the prior year through November of the current year. The Census Bureau must adjust the ACS data to make them comparable.[3] These and other differences described in the Czajka paper add up to differences in the survey estimates of income and thus poverty. In some respects, the results are fairly close. Czajka points out that aggregate income ranges from $6.12 trillion in the NHIS to $6.47 trillion in the CPS—a spread of just 5 percent. This means that in a single question the NHIS captures 95 percent as much total income as the CPS, and the ACS, with a simple instrument filled out primarily by respondents rather than a trained interviewer, captures 98 percent as much total income as the CPS.

However, Czajka makes the point that distributions are important, and there are important differences when income collected by these three

[3] To convert the ACS income to a common reference period—specifically, the calendar year in which the data were collected—the Census Bureau applies an inflation adjustment, defined as the ratio of the average monthly price index for the survey year to the average index for the reference period. These monthly adjustment factors are used internally and are applied to published estimates, but the ACS public-use file contains only an average adjustment factor for the 12 survey months, because the survey month is not reported. To calculate income relative to poverty, the Census Bureau adjusts the poverty thresholds rather than the reported income. That is, the income reported for a given reference period is divided by the average monthly threshold for that reference period. The estimates of income relative to poverty derived in this manner are reported in the public-use file.

surveys is examined by quintile. Although the ACS aggregates are within a percentage point of the CPS aggregates (both above and below) through the first three quintiles, they drop to 98 and 97 percent of the CPS in the fourth and fifth quintiles, respectively. The NHIS captures only 85 percent as much total income as the CPS (and the ACS) in the bottom quintile.

The differences show up in the poverty rates produced by each of the surveys. The CPS and the ACS yield poverty rates of 12.2 percent and 12.5 percent, respectively, but the NHIS estimates a somewhat higher poverty rate—14.7 percent. Likewise, estimates of children in low-income families from the CPS and the ACS lie very close to each other, whereas the NHIS finds somewhat more. Interestingly, for purposes of comparing the number of children in families likely to be eligible for CHIP and improved benefits under the Patient Protection and Affordable Care Act, the estimates of near-poor children vary from 14.9 to 15.4 million or 21.1 to 21.5 percent across the three surveys.

In his paper and presentation, Czajka drew out of these comparisons several implications of using the different surveys as a source of income and poverty data. Despite differences in measurement, "the ACS produces estimates of income and health insurance coverage that look strikingly similar to those obtained from the CPS" (see Chapter 10), giving support to the use of the ACS to develop direct estimates of low-income uninsured children at the state level and even lower levels of geography. As income is disaggregated, however, some consequential differences emerge owing to collection methods and reference periods. The longer reference period in the ACS means that it tends to lag the CPS in response to changes in the economy, an important issue when assessing the change in health insurance coverage of low-income children in response to the business cycle.

Finally, when it comes to income data, nonresponse and rounding are important considerations. Czajka stated that for both the CPS and the NHIS, about a third of total income is imputed. The CPS imputes a "staggering" figure of 63 percent of the dollars for asset income. The allocation rate for the ACS is only about half of what it is for the CPS. Rounding is important when comparing the survey results. Czajka illustrates the impact of rounding in work that found that close to one-third of the ACS earnings responses are exactly divisible by $5,000. This has important implications when using the data to simulate program eligibility because the data are artificially affected by cutoff points.

SUMMARY

Several observations with regard to the major federal surveys were made by the presenters and are summarized here. Perhaps the most significant is that differences in the estimates of health insurance coverage among the three major current sources are not analytically significant. The CPS, the ACS, and the NHIS all produced coverage estimates that rounded to 85 percent in 2008.

The choice of which data source to use largely depends on the question that is asked, since they have different features related to sample size, level of detail used in questions collecting information about coverage, other subjects asked about, characteristics of the interview, and postcollection processing that affect their estimates of health insurance coverage. For example, if the need is for a point-in-time estimate, the user would turn to the ACS, whereas the CPS yields an all-year view of coverage. Each of the surveys has limitations when it comes to monitoring coverage, so Kenney and Lynch suggest that it is important to benchmark key estimates to other surveys.

Some sophisticated analysis by the State Health Access Data Assistance Center, using an "enhanced" CPS series, gives an approximation of what future estimates might look like. Turner reported that the 2008 estimates showed an ACS estimate of 14.6 percent uncovered, which was nearly identical to the NHIS 14.8 percent and actually identical to the CPS rate. On the basis of this work, she observed that the ACS estimates of health insurance coverage looked reasonable, and, when the logical edits are implemented for 2009, it is possible that the ACS estimate of uninsured may be nearly identical to the "best" NHIS estimate for overall coverage.

Much work needs to be done to improve the precision of the survey-based estimates of uninsured children, including modifying questionnaires, enhancing content, and expanding what is available in public-use files, improving clarity of published estimates, improving documentation, giving data users more information about reasons for differences in survey estimates, expanding state-level data on access and service use, and editing cases with misreported coverage. Many of the presenters emphasized conducting targeted methodological research, building bridges between the surveys so they could benefit from the strengths of one another, and providing data users more information for analyzing and possibly further adjusting data.

4

Administrative Databases

Administrative data collected in the course of program operations, no matter how good and complete, are unlikely to be able to fully substitute for survey data in that, although they can depict the population enrolled and served in public programs, they cannot delineate the population that is truly uncovered. An uncovered population as delineated by program data consists of children who may well have coverage through other (mostly private) programs and those who have fallen through the cracks and have not been enrolled in the programs.

Nonetheless, the counts of actual children enrolled and served that are the product of the collection of program operating data are far better counts of the served population than can be obtained through surveys, for the reasons given in Chapter 3. This chapter discusses the major sources of administrative data on children covered in public programs, assesses their strengths and weaknesses, and addresses steps being taken by the federal and state agencies responsible for collecting and compiling the data to improve their quality, timeliness, and representativeness over time.

MAJOR ADMINISTRATIVE DATABASES

The three major administrative databases maintained by the federal government are the CMS-64 Quarterly Expense Report to the Centers for Medicare & Medicaid Services (CMS), the Medicaid Statistical Information System (MSIS), and the Children's Health Insurance Program (CHIP) Statistical Enrollment Data System (SEDS). Each of these is discussed below.

CMS-64 Quarterly Expense Report

Form CMS-64 is a statement of expenditures for which states are entitled to federal reimbursement under Title XIX of the Social Security Act. It reconciles the monetary advance made on the basis of Form CMS-37, filed previously for the same quarter. Consequently, the amount claimed on Form CMS-64 is a summary of expenditures derived from source documents, such as invoices, cost reports, and eligibility records. At the workshop, David Rousseau made the point that the CMS-64 is primarily collected to reconcile payments to states from the federal treasury for services delivered to Medicaid beneficiaries. It has a great deal of information about overall spending but not very much about what is driving that growth, because, he pointed out, there is no information on utilization, enrollment, or the types of beneficiaries that use that service, and managed care spending is not separated into the services.

Medicaid Statistical Information System

The purpose of the Medcaid Statistical Information System (MSIS) is to collect, manage, analyze, and disseminate information on eligibles, beneficiaries, utilization, and payment for services covered by state Medicaid programs (see www.cms.gov/MSIS [September 2010]). The MSIS data serve multiple purposes, including health care research and evaluation activities, program utilization and expenditures forecasting, analyses of policy alternatives, responses to congressional inquiries, and matches to other health-related databases. States provide CMS with quarterly computer files containing, among other data items, specified data elements for persons covered by Medicaid (eligible files). If a person is covered by Medicaid (or CHIP) for at least 1 day during the reporting quarter, their demographic and monthly enrollment data resides in this file. Claims records contain information on the types of services provided, providers of services, service dates, costs, types of reimbursement, and epidemiological variables. The data files are subjected to quality assurance edits to ensure that the data are within acceptable error tolerances, and a distributional review verifies the reasonableness of the data. Once accepted, valid tape files are created that serve as the historical source of detailed Medicaid eligibility and paid claims data maintained by CMS.

Statistical Enrollment Data System

States submit quarterly and annual CHIP statistical data to CMS through the SEDS automated reporting system (U.S. Department of Health and Human Services, Centers for Medicare & Medicaid Services, 2010, p. 47). Using forms provided by CMS, states report unduplicated counts of the number of children under age 19 who are enrolled in separate CHIPs and Medicaid expansion CHIPs.

ASSESSMENT OF THE MAJOR ADMINISTRATIVE DATABASES

Rousseau pointed out that the databases do a good job of measuring what they are intended to measure. For example, the measures of program enrollment are detailed and extensive. Year-to-year enrollment is the primary driver of Medicaid spending growth and is, in fact, the only driver of Medicaid's increasing share of overall national health care spending. The data systems permit more detailed analysis of enrollment data. But there are issues with timeliness and completeness, which has driven organizations such as the Kaiser Family Foundation to collect Medicaid and CHIP administrative data directly from the states to track enrollment growth on the same time line as spending data from CMS become available, mostly because enrollment data lag spending data by at least a year. Kaiser collects administrative data directly from all 50 states and the District of Columbia in partnership with researchers at Health Management Associates. The organization finds that having enrollment data from the same period as the latest spending data allows it to analyze spending trends far sooner.

Depending on the questions that are asked of the administrative data, the sources change and the answers differ in ways that may be confusing. For example, Medicaid and Medicaid expansion CHIP are reported in MSIS. If the data user is looking for the population that is "ever enrolled by fiscal year," the source is MSIS Medicaid State Summary Mart; the population "ever enrolled in a calendar year" comes from the Medicaid Analytic eXtract (MAX), a special repository of research data derived from MSIS; and the population "enrolled at a point-in-time" is also available from MAX (enrolled in any given month). The "ever enrolled" and "point-in-time" estimates are themselves quite different: the "ever enrolled" counts are inclusive annual (fiscal or calendar year) counts of Medicaid and CHIP enrollees, which include those enrolled for any time, from a single day up to a full year, and "point-in-time" estimates are the count as of a single, specified day. These files may or may not represent the state CHIP programs, since stand-alone state CHIP enrollments are reported at state option. A total of 44 states have either combination (Medicaid and stand-alone) or stand-alone only CHIPs, but 21 do not report all stand-alone CHIP enrollees in MSIS, so users have to look elsewhere to get a total for both CHIP elements. This is an important issue, because MSIS includes all Medicaid and Medicaid expansion CHIP enrollments (M-CHIP), but enrollment in the state CHIP (S-CHIP), which was more than two-thirds (69 percent) of enrollment in 2008, is not consistently included in the MSIS. This limits the ability to look at children's enrollments in Medicaid and CHIP in a comparable way.

There are also troubling differences in the totals that are obtained from the different administrative data sources. For example, Rousseau found that, after adjusting for differences in scope and design, since the late 1990s, MSIS spending is consistently below CMS-64 spending totals. Differences between the CMS-64 and the MSIS by state range from 45 to –13 percent, with six states reporting more than 10 percent higher spending in the MSIS than in the CMS-64, which potentially means they are losing out on some federal dollars. In total, the differences range from about $5 billion to more than $16 billion per year. These differences are noted even after excluding spending from the CMS-64 that is not intended to be put into the MSIS, such as administration or a disproportionate share of hospital spending or accounting adjustments.

The differences between administrative data counts and survey estimates are sometimes explainable. Rousseau mentioned that one difference between the National Health Interview Survey (NHIS) and MSIS enrollment data for the same time period is that the MSIS includes retroactive eligibility—persons who were deemed not eligible at the point in time but who have become eligible over time. Another difference is in definitions, a common difference being varying age cutoffs to describe the population.

A summary of the strengths and weaknesses of the three main administrative databases is found in Table 4-1.

STEPS BEING TAKEN TO IMPROVE THE ADMINISTRATIVE DATABASES

David Baugh summarized the intensive program under way at CMS to rationalize and improve these administrative databases, particularly the MSIS data. These efforts, according to Rousseau, are bearing fruit, in that MSIS data quality continues to improve, with far fewer beneficiaries of unknown eligibility and far less spending on unknown services.

Baugh discussed an initiative called the Medicaid and CHIP Business Information and Solutions (MACBIS) council, which provides leadership and guidance in support of efforts toward a more robust and comprehensive information management strategy for Medicaid, CHIP, and state health programs. The MACBIS is charged with promoting consistent leadership on key challenges facing state health programs, including quality, access, value, and integrity; improving the efficiency and effectiveness of the federal-state partnership; making data on the Medicaid, CHIP, and state health programs more widely available to stakeholders; and reducing duplicative efforts within CMS and thus minimizing the burden on states. Under the auspices of MACBIS, the agency is also developing an

TABLE 4-1 Strengths and Weaknesses of Primary CMS Administrative Databases for CHIP

Strengths	Weaknesses
CMS-64	
• Relatively current information (2009 has been available since early 2010) • Mature reporting system, tied to payment, helping to ensure data quality at both state and federal levels	• Tracks only aggregate spending by service (no utilization or enrollment) • Managed care spending not separated into services • Not intended for research; understanding trends often requires following up with states regarding their submissions (e.g., Arizona and Vermont)
Medicaid Statistical Information System (MSIS)	
• Granular data, supporting a wide range of spending and enrollment analyses • Has improved markedly over the years, both in terms of timeliness and data quality	• Size and complexity make it susceptible to submission delays, although much improvement has been made (2008 nearly complete) • Its very size and complexity can contribute to missing data and a wide range of anomalies • Managed care spending not separated into services
Statistical Enrollment Data System	
• Available earlier than MSIS	• Collects only aggregate enrollment data

SOURCE: Data from Rousseau (2010).

enhancement to the MSIS known as MSIS Plus, which expands reporting to support anti-waste, fraud, and abuse programs and other activities.

New funding has been made available to improve the MSIS data through an expansion of MAX data activities funded by the American Recovery and Reinvestment Act (ARRA). CMS also coordinates a project known as the Medicaid and CHIP Fee-for-Service, Managed Care, Eligibility, and Provider Data Improvement Project, which is identifying the data elements that CMS needs in order to manage Medicaid, CHIP, and other state programs.

Another initiative is to improve the identification of program enrollees by obtaining identification crosswalks from Medicaid agencies. This work has been under way for some time, but to date, corrections have been implemented in MAX but not MSIS. The agency has also turned attention to verification of Social Security numbers on a limited basis using the

Social Security Administration High Group List. CMS is exploring more complete verification in concert with the Social Security Administration, the Census Bureau, and the states.

Duplication in the count of enrollees across the states and over time is an issue for CMS and data users. Baugh estimated that about 1 million of the 60 million Medicaid/CHIP enrollees are duplicates. The duplicates will be reduced by creation of a MAX Enrollee Master file (MAXEM), on which work is now under way.

Lastly, he reported that under the umbrella of the MAX expansion project, funded by ARRA, there is increased coordination between CMS and State Technical Advisory Groups to describe CHIPs, identify their issues, challenges, and concerns, and develop criteria for selecting volunteers. As part of this initiative, CMS is working to develop state-specific technical assistance plans in 10 to 15 pilot states, which include regular communications and working closely with data task teams, with the objective of improving MSIS reporting. However, at this time, state participation in these quality improvement programs is voluntary, regulations have yet to be promulgated, and states have limited resources to make improvements that have been identified.

SUMMARY

In summarizing his views on the status of administrative records for estimating children's health insurance coverage, Rousseau observed that administrative data are a good resource for research and benchmarking, but they are not a perfect gold standard. For one thing, the data are collected for other purposes (e.g., CMS-64) and not for specific research questions, so they may not be able to fully illuminate some key issues, such as the lack of managed care information in MSIS. However, they are increasingly available and researchers can and do use them for a variety of purposes. Despite their limitations, administrative data complement survey data and are a critical component in all work of CHIP and Medicaid researchers.

Administrative data are critical for state-level analysis of subpopulations for which surveys are often limited by sample size and other methodological issues. Real-time administrative data on enrollment are critical to decision makers and analysts. CMS has made great strides in improving the availability and timeliness of administrative data, but more work needs to be done to ensure data quality and make enrollment data contemporary with spending. The new health reform legislation increases the need for timely, high-quality administrative and survey data.

5

State Data Collections

From the beginning of the Children's Health Insurance Program (CHIP), state agencies have had considerable flexibility in designing and administering the programs, and they have a mandate under the legislation to evaluate their programs. Indeed, state agencies have a multitude of responsibilities that require access to information in order to efficiently manage the programs and assess progress toward the goal of ensuring full coverage for eligible children.

This chapter discusses the motivations for state-specific surveys and summarizes information about the extent and design of such surveys. Examples are presented of the experience in two states, one with an extensive program of state-sponsored surveys and another that relies entirely on federal surveys.

MOTIVATION FOR STATE SURVEYS

Not all states conduct state surveys, but all states do have Medicaid and CHIP responsibilities that require the kind of coverage and program management information that is collected in such surveys. In her presentation to the workshop, Lynn Blewett catalogued the various responsibilities of state agencies that create a need for adequate, timely, and accessible CHIP data.

The growing and currently most critical role of the states is to facilitate implementation of access provisions in health reform. This requires information that supports programs to expand Medicaid, implement state insurance exchange and regulation, provide for possible public plan implementation at the state level, and to implement insurance regulations, including young adult dependent coverage.

A continuing responsibility for states is to meet CHIP reporting requirements, which are largely established by the Centers for Medicare & Medicaid Services (CMS). These constitute a series of annual state reports to CMS on progress in reducing the number of uninsured children, to be submitted in specified formats with data elements established in the CMS reporting programs (see Chapter 4). The requirements of the CHIP challenge, discussed in Chapter 2, now require updates with estimates of coverage on a semiannual basis.

States also have a need to effectively target outreach, enrollment, and safety net strategies. Information is needed on insurance status, age, substate geographic location, income, and race/ethnicity in order to guide and assess these activities.

And states need to prepare budgets, and, to properly accomplish the budgeting function, they need to be able to forecast activities. For the most part, inputs to forecasting models are based on expansion or contraction activities. As part of the budget process, states need to prepare and justify distribution formulas for state funds to localities, Blewett observed.

She suggested that, in order to accomplish the various responsibilities outlined above, a set series of state data requirements has evolved, generally characterized by:

1. a state representative sample;
2. a large enough sample and a sample frame that provides for reliable estimates for subpopulations including low-income children, race/ethnic groups, and geographic areas, such as county or region;
3. timely release of data, including tabulated estimates of health insurance coverage released within a year of data collection; and
4. access to microdata through readily available public-use files with state identifiers to allow states do conduct their own analysis and policy simulations.

COORDINATED STATE COVERAGE SURVEY

For the reasons given in Chapters 3 and 4, the current federal surveys and national administrative databases have not been judged sufficient to fulfill these data requirements, so states have for some time sponsored their own ongoing surveys. Blewett reported that 19 states have used an instrument called the Coordinated State Coverage Survey (CSCS), which was developed for public use by the State Health Access Data Assistance Center at the University of Minnesota.

The CSCS is a household telephone survey designed to monitor the uninsured and provide reliable state-level insurance coverage estimates to inform state health policies. It contains a core of questions on health insurance coverage, demographics, and access to care. Some states have supplemented the survey to provide locally important data. Topics that have been added by supplementing the survey include willingness to pay (Indiana, in 2003), loss of coverage (Alabama, in 2003; Oklahoma, in 2004), dental coverage (Alabama, in 2003), attitudes toward and awareness of state reforms (Massachusetts, in 2008), and Medicare supplemental policies (Minnesota, in 2004 and 2007).

Most of the telephone surveys are conducted in a random digit dialing (RDD) mode, although several lead states are moving to dual frame surveys with cell phone samples. It has been found that, by 2009, 24.5 percent of households had cell phone–only telephone access. These cell phone–only households are significantly more likely to lack health insurance compared with households with land line telephone service. Although in the past, most states have been able to account for coverage error involving households without phone service through a weighting adjustment, it has not been possible to successfully make similar adjustments for cell phone–only households.

There are growing concerns about these state surveys. Like many other surveys, state survey managers have a concern about an overall trend toward declining response rates on the RDD surveys. The surveys are also quite costly. Blewett said that states commit between $250,000 and $500,000 on average for these surveys, with financing usually cobbled together from the state general fund, CHIP evaluation federal administrative match, conversion, and contributions from foundations.

GROWING ACCEPTANCE OF THE AMERICAN COMMUNITY SURVEY

Although it can be assumed that states would be eager to forgo the difficulty and cost of conducting their own state survey, they have been wary of placing their trust in the federal surveys. This attitude may be changing. Since the American Community Survey (ACS) added a question on health insurance coverage and has matured and gained the credibility that comes from increasing reliability for smaller geographic areas, it is becoming a more accepted source of health insurance coverage information at the state and substate level.

Blewett summarized the state perspective on the ACS compared with the Current Population Survey (CPS). On the positive side, the ACS has a large sample size, geographic coverage, annual estimates, and public-use files. Importantly, imputation, when necessary, is done in each state independently. The questions address point-in-time coverage and are asked of each person in the household.

It is also becoming obvious that the ACS comes out very well in a quality comparison with both the CPS and the state surveys. As shown in Figure 5-1, the relative standard errors for the uninsurance question are smaller for the ACS in nearly every state, and they are significantly smaller in some states.

On the negative side, the ACS does not have the ability to insert state CHIP names, as the CPS now is able to do. Blewett indicated that the states need the ability to add state program names, such as Badger Care and Minnesota Care, since participants relate to those program names and may not respond correctly to a vague program description; it is a matter of accuracy, she said.

Some of the results of the ACS, juxtaposed with the prior estimates from the CPS, are troublesome to explain. In some states, the CPS rates are higher; in others, lower, and there does not seem to be a pattern to help explain the differences. It may be that the ACS data are being questioned just because they are new, different estimates and unfamiliar to audiences that are used to data from the CPS and state surveys. In addition, the ACS has no additional health status or access questions, so its ability to fully substitute for state surveys is limited.

NATIONAL HEALTH INTERVIEW SURVEY AS A SOURCE OF STATE DATA

Blewett also discussed a fourth source of state-level information—the National Health Information Survey (NHIS). The NHIS publishes health insurance coverage estimates for 20 selected large states each year. It also has the advantage of early availability of data compared with other data sources. For example, the calendar year 2009 data were available in June 2010.

But the NHIS has shortcomings as well. It does not publish data for 30 states, so geographic coverage is limited. Also, the smallest geographic identifier available on public-use microdata from the NHIS is the census region, which tends to limit state-level or subpopulation analysis. The rules placed on users of the public-use microdata mean that those wanting state-level analysis must obtain access to state identifiers in National Center for Health Statistics Research Data Centers. Although the NHIS is a rich health-related data source, with a good point-in-time health insurance question, it has limited state-level use.

As a result of the various limitations of the data available at the state level, Blewett contends, it is difficult for state analysts to pick one mea-

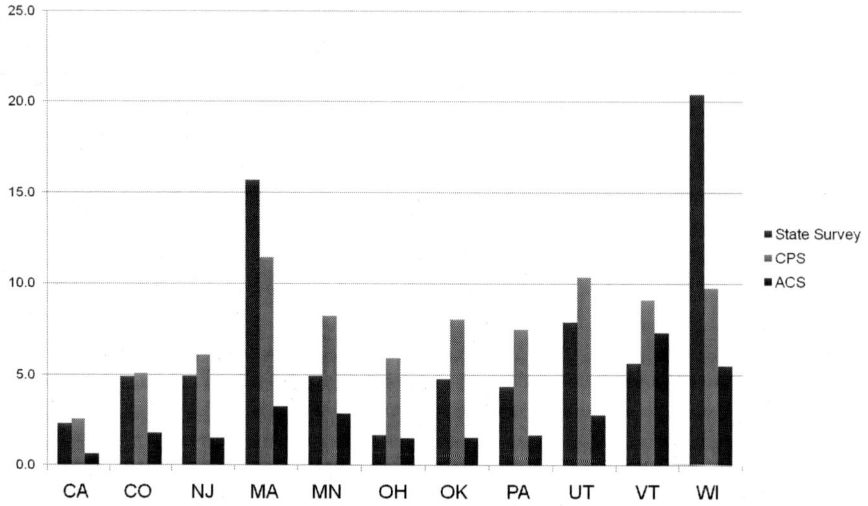

FIGURE 5-1 Comparison of relative standard errors for uninsurance by survey source, all ages.
NOTES: 2009 CPS-ASEC and 2008 ACS data are from public-use data. State survey data are from published reports and personal communication. Relative standard errors are defined as the SE divided by its mean.
SOURCE: Blewett (2010).

sure and stick to it. She observed that the more data-sophisticated states (about one-third of all states) use different surveys for different purposes. Although the CPS now provides the primary estimate for uninsurance when comparing across states, states with their own surveys tend to use it internally, and state policy makers have been educated on the differences across surveys. Even though the ACS is gaining popularity, many states are expected to continue to sponsor state surveys. Their policy decisions are likely to continue to rest with state survey data, for the reasons of familiarity (states are used to using state-specific data to inform policy decisions) and such features as the ability to add questions quickly that might be useful for answering state and national policy questions.

THE VIRGINIA EXPERIENCE

John McInerney provided context on the use of data in a typical state by summarizing the experience with CHIP in the Commonwealth of Virginia. Like other states, Virginia has found that managing CHIP has been a learning experience. The initial CHIP was judged to be unsuccessful—it incorporated a 12-month waiting period and covered children in families with incomes only up to 185 percent of the federal poverty level (FPL). As a consequence, the program was redesigned in 2001 to include 12-month continuous eligibility, coverage offered up to 200 percent FPL, a reduced waiting period, and a name change to FAMIS (Family Access to Medical Insurance Security). Initially, the change was successful.

As FAMIS grew, the rate of uninsured children decreased in Virginia between 2000 and 2004. However, as measured by survey data, that progress was reversed between 2004 and 2006. Incredibly, FAMIS enrollment continued to grow, but an increase in uninsured children was registered in survey data. The uninsurance rate measured by the CPS and the ACS continued to be high in 2007 but saw a slight decline in 2008, consistent with a modest increase in the FAMIS/Medicaid enrollment in that year but inconsistent with the fact that 2008 was a year of rising unemployment. McInerney pointed out that these apparent anomalies have been difficult to explain to nontechnical audiences.

As a state that has only rarely produced its own survey (the last state survey was in 2004), Virginia analysts and advocates will continue to rely on national survey data for state-level analysis. Analysts are looking forward to the increasing availability of county-level data that will come with the ACS, since Virginia has different economic conditions by region, and foundations and direct service providers want more localized data. They have not been able to get much access to county counts in Virginia because state privacy policies constrain the state from releasing county enrollment data for Medicaid and FAMIS.

THE MASSACHUSETTS EXPERIENCE

Unlike Virginia, the Commonwealth of Massachusetts has heavily invested in sponsoring state-specific surveys. In her presentation, Sharon Long discussed two state-specific surveys in Massachusetts—the Massachusetts Health Insurance Survey (MHIS) and the Massachusetts Health Reform Survey (MHRS). She compared estimates of uninsurance rates for Massachusetts across state-specific surveys and national surveys and summarized lessons learned from the two Massachusetts surveys.

The motivations for the Massachusetts surveys were similar to those reported by Blewett for states that have elected to sponsor their own surveys. One motivation was the potential for larger state sample sizes, both overall and for key population subgroups and substate geographical areas, so details not available from the national surveys could be produced. Likewise, in embarking on an ambitious change in health care policy, there was a need for information on Massachusetts-specific insurance and health care programs. The needed information included information on insurance coverage, but it extended beyond that to information about health care access and use, costs, quality, barriers to care, awareness of reform, and attitudes toward the reforms being initiated. The health care agencies in Massachusetts also needed more timely access to data to track reform and inform policy and program design than was possible with national surveys. Massachusetts judged that the ACS, which provides a much larger sample size for Massachusetts than is available from any other survey, does not address the other needs—state-specific insurance coverage options in the survey questions, information on health care outcomes beyond insurance coverage, and more timely data.

Massachusetts Health Insurance Survey

The MHIS is sponsored by the Massachusetts Division of Health Care Finance and Policy. It began about the time that CHIP was initiated in 1998 and was redesigned in 2008 to expand the survey sample frame to include all residential households (not just those with a land line telephone) and to modify the questionnaire to collect more of the health insurance and health care options in the state in response to health care reform changes.

The current survey consists of a sophisticated dual sample frame to capture cell phone–only households with both RDD telephone and address-based (AB) samples. There are multiple survey modes—mail, telephone, and web. The survey questionnaire is in three languages—English, Spanish, and Portuguese. It goes to 4,000 to 5,000 households each year, collecting data on health insurance coverage, health care access and use, health care costs, and attitudes toward health reform.

Long reported that external data sources are used to obtain as much contact information as possible for addresses in the RDD sample and phone numbers for the AB sample. Similarly, the survey offers multiple survey modes to as many as possible—mail, telephone (call in and call out), and web. The design is summarized in Table 5-1.

Massachusetts Health Reform Survey

The Massachusetts Health Reform Survey is sponsored by the Blue Cross/Blue Shield of Massachusetts Foundation, with support in initial years from the Commonwealth Fund and the Robert Wood Johnson Foundation. It was initiated in 2006 in an effort to track the state's health reform initiative. The design is an RDD telephone sample, with questionnaires like the MHIS in three languages—English, Spanish, and Portuguese. Fielded in the fall of each year, the survey sample is 3,000 to 4,000 nonelderly adults each year. Some groups (low- and moderate-income adults and uninsured adults) are oversampled. It provides information on

TABLE 5-1 Characteristics of the Random Digit Dialing and Address-Based Samples by Mode

	Random Digit Dialing Sample		Address-Based Sample	
Mode	With Known Address (43%)	Without Known Address (57%)	With Known Phone Number (83%)	Without Known Phone Number (17%)
Web	x		x	x
Mail	x		x	x
Call in to a toll-free number	x		x	x
Outbound call by the survey firm	x	x	x	

SOURCE: Long (Chapter 12, in this volume).

health insurance coverage, health care access and use, health care costs, attitudes toward health reform, and health plans.

Long compared the results of the state surveys with the CPS, the NHIS, and the Behavioral Risk Factor Surveillance System[1] in 2006, prior to the passage of health care reform and the redesign of the MHIS, and concluded that each produced a very different estimate of the uninsurance rate for Massachusetts. The three surveys that provided estimates for children (CPS, NHIS, and MHIS) had uninsurance estimates that ranged from 2.5 percent (MHIS) to 7.1 percent (CPS). These differences were traced to several factors, such as differences in the sample populations included in the surveys, differences in the wording of the insurance questions asked in the surveys, differences in question placement and context, differences in survey design and fielding strategies, and the survey time frame. Importantly, the surveys conducted in 2008, after health reform, showed low rates of uninsurance among children, with the CPS estimating that 3.4 percent of Massachusetts children were uninsured in 2008, compared with an estimate of 2.1 percent in the ACS and 1.2 percent in the MHIS.

Long drew several lessons from the rather extensive Massachusetts survey experience. She observed that state surveys have been essential for timely feedback on the impacts of health reform in a period of rapid state and national policy development. Yet both of the Massachusetts surveys

[1] The system covers people aged 18-65, so it is generally not useful for directly estimating children's health insurance coverage.

have faced challenges. It is difficult to maintain funding over time, which means, for example, that it was necessary to make decisions on survey design with little research on survey methods because of shortage of funds to support that basic research. In the end, Massachusetts obtained useful information, but it was in isolation. Analysts need comparisons with other states for context and to support stronger evaluation designs.

Despite the relatively successful experience with state surveys in Massachusetts, Long contended that federal surveys are still needed. She made several recommendations of ways to make federal survey data more useful: (a) provide much larger state and local area samples, both overall and for key population groups (including children); (b) make state identifiers available outside of research data center settings; (c) add more geocoding of state and local areas; (d) for the ACS, add state program names to health insurance questions; (e) expand survey content to include questions on access, use, and costs of care, along with other issues of relevance to national health reform; and (f) make data files available more quickly and in user-friendly formats to facilitate their use by state analysts.

6

Modeling Strategies for Improving Estimates

At the workshop, the participants were reminded that, in a 2003 paper discussing the issue of variability in the Current Population Survey (CPS) and the American Community Survey (ACS), the State Health Access Data Assistance Center (SHADAC) suggested that the "most desirable path would be to create estimates via statistical modeling that combine the strengths of the CPS-ACS and the fully implemented ACS. This method would be similar to the one used by the Census Bureau to allocate Title I education funds" (State Health Access Data Assistance Center, 2003). Although this suggestion has not been implemented, it indeed reflects a growing recognition that models for the generation of estimates, using survey and administrative data as inputs, could well have positive attributes for the equitable distribution of program funds to subnational geographic areas. The Kenney and Lynch paper (Chapter 8, in this volume), discussed in Chapter 3, suggested that it is possible that model-based estimates that combine ACS and CPS Annual Social and Economic Supplement (ASEC) estimates with administrative records data could be an improvement over single-source survey data.

This chapter summarizes presentations on the two main Census Bureau models—the Small Area Income and Poverty Estimation project (SAIPE) and the Small Area Health Insurance Estimation project—that currently produce estimates for small areas. It also reports on another approach that is gaining traction in the statistical community: the combination of data from multiple surveys.

THE SMALL AREA INCOME AND POVERTY ESTIMATION MODEL

The SAIPE project is an ongoing Census Bureau program to estimate numbers of low-income school-age children by state, county, and ultimately school district, based on data from the CPS, tax information, the Food Stamp Program, and the latest decennial census. SAIPE has a long history. It has been the subject of extensive development at the Census Bureau and evaluation by a previous National Research Council panel. The SAIPE approach to county-level estimates was developed in response to legislation in 1994 (National Research Council, 2000, p. 3) calling for the Census Bureau to supply "updated estimates" of county-level child poverty for use in allocating Title I education funds to counties in 1997-1998 and 1998-1999, and thereafter to provide estimates at the school district level.

William Bell pointed out at the workshop that the SAIPE was designed to respond to the prototypical small-area estimation problem that although some surveys produce reliable estimates at national level, subnational (e.g., state or county) estimates are desired, and the sample is not large enough to support the small-area estimates. The idea is to apply a statistical model to "borrow information" across areas and from other data sources to improve estimates. The key features in a successful model are the other data sources used, the form and underlying assumptions of the model, and diagnostics that can be used to check model assumptions.

The SAIPE model follows this form, using as data sources sample surveys, administrative records, and census data. Except for a complete enumeration from the census, all of these sources have errors of one of three types: (1) sampling error, that is, the difference between the estimate from the sample and what is obtained from a complete enumeration done in the same way; (2) nonsampling error, that is, the difference between what is obtained from a complete enumeration and a population characteristic of interest ("desired target"); and (3) target error, that is, the difference between what a data source is estimating (its target) and the desired target. As an example of target error, Bell pointed out that the model uses food stamp participants to measure poverty, but Food Stamp Program participant data are only an approximation of the desired target.

Bell summarized the strengths and weaknesses of the main data sources for SAIPE purposes. The CPS, the primary national survey measuring population and poverty each year, provides the SAIPE program with national county-level child poverty estimates, through sample-weighted estimates of the numbers and the proportion of poor children among children aged 5-17 related to the primary householder (poor related school-age children). Since 2005, these data have been obtained from the ACS. Both the CPS and the newer ACS produce direct poverty estimates that are up to date but that have large sampling errors for "small" areas. The administrative predictors are the county numbers of child tax exemptions for families in poverty and of all child exemptions reported on tax returns, along with county numbers of households participating in the Food Stamp Program. Although there is no sampling error in these administrative data sources, there is target error, in that the data are not collected specifically to measure poverty.

The SAIPE uses a basic univariate model developed by Bob Fay and Roger Herriott in 1979. In this model,

$$y_i = Y_i + e_{ii} = (x_i'\beta + u_i) + e_i,$$

given:

y_i = direct survey estimate of population quantity Y_i for area i,
x_i' = vector of regression variables for area i,
β = vector of regression parameters,
u_i = area i random effect (model error) ~ i.i.d. $N(\mathbf{0},\sigma_u^2)$, and independent of e_i, and
e_i = survey errors ~ ind. N($\mathbf{0}$,v_i) with v_i assumed known.

In application, the SAIPE produces a state-level poverty rate for children aged 5-17. The direct estimates y_i were originally from the CPS, but since 2005 they have been taken from the ACS. The regression variables in x_i' include a constant term and, for each state, a pseudo state child poverty rate from tax return information, a tax "nonfiler rate," a Food Stamp Program participation rate, and a state poverty rate for children aged 5-17 estimated from the previous census, or residuals from regressing previous census estimates on other elements of x_i' for the census year.

The actual model is estimated as follows:

- Given σ_u^2 and the v_i, let $\Sigma = \text{diag}(\sigma_u^2 + v_i)$, estimate β by WLS:

$$\hat{\beta} = (X'\Sigma^{-1}X)^{-1}X'\Sigma^{-1}\boldsymbol{y}$$

where y = $(y_{1,\ldots,} ym)'$, and X is $m \times r$ with rows x_i'.

- Given, v_i, estimate σ_u^2 by the method of moments, ML, REML, or do so via the Bayesian approach.
- Use the standardized residuals: $r_i = (y_i - x_i'\hat{\beta}) / [\text{Var}(y_i - x_i'\hat{\beta})]^t (.5)$ for model diagnostics.

A technique known as best linear unbiased prediction is used to improve the variance. In this case:

$$\hat{Y}_i = h_i y_i + (1 - h_i) x_i'\hat{\beta},$$

where $h_i = \sigma_u^2 / (\sigma_u^2 + v_i)$.

One can also compute $\mathrm{Var}(Y_i \mid \mathrm{y, model}) = \mathrm{Var}(Y_i - \hat{Y}_i)$. The model assumes $E(Y_i) = E(y_i) = x_i'\beta$, and the WLS fitting then implies that $E(x_1 i^{1r}\beta'') = x_i'\beta$.

Bell pointed out that, under this assumption, the model fitting linearly adjusts the x_{ij} so that $x_i'\hat{\beta}$ estimates Y_i. For example, if there is only one regression variable, x_i, plus an intercept, then

$$x_i'\hat{\beta} = \hat{\beta}_0 = \hat{\beta}_1 x_i,$$

which is a linear adjustment of x_i to estimate Y_i.

Variance improvement from modeling and the best linear unbiased prediction is reflected by

$$\%\ \mathrm{diff} = 100 \times = \frac{Var(Y_i \mid y, \mathrm{model}) - v_i}{v_i}.$$

Negative values of % diff represent improvements.

Bell observed that the work on the SAIPE model indicates that small-area models can reduce variances of direct survey estimates for areas with small samples (and provide estimates for areas with no sample). In his presentation, he used examples from California, Indiana, and Mississippi, which showed that the model improved variances but the improvements were much more substantial for the two smaller states (50-60 percent) than California (26 percent). He also showed that the model estimates generally ignore nonsampling error in the direct survey estimates that provide the dependent variable in the model, which implicitly defines the target to be what the survey is estimating. He said that model fitting effectively adjusts the independent variables so that model predictions estimate the target defined by the dependent variable and that success in modeling depends on the quality of the data sources and the appropriateness of the model.

Another positive virtue of modeling is that various diagnostics (plots and tables) can be used to check the model. Bell referred to a National Research Council report (2000) on SAIPE that suggested a number of diagnostics to be applied to the models. These diagnostics were based on standardized residuals from the model plotted against various geographic areas and population groups to reveal patterns. In one test, for example, the Western census region showed negative residuals over a number of years, suggesting the need to modify the model.

The lessons learned in the development and testing of the SAIPE model were carried over into the development of a model for health insurance in small areas.

SMALL AREA HEALTH INSURANCE ESTIMATION MODEL

The Small Area Health Insurance Estimation (SAHIE) model is an outgrowth of the SAIPE model, although it goes beyond SAIPE in that it estimates the numbers in income groups and then, within income groups, the number of uninsured.

Small-Area Health Insurance Estimation Coverage

At the workshop, Mark Bauder presented a paper coauthored by Brett O'Hara on the SAHIE model. He reported that SAHIE began on an experimental basis in 2005 in order to meet needs of the Centers for Disease Control and Prevention to estimate the population of low-income uninsured women in particular age groups. The first live estimates were published in 2007, and the SAHIE program currently publishes state health insurance coverage estimates by age, sex, race, Hispanic origin (i.e., demographic characteristics), and income categories (0-200 percent *and* 0-250 percent of the poverty threshold and the total poverty universe). For counties, SAHIE produces estimates for fewer items—age, sex, and income categories (0-200 percent *or* 0-250 percent of the poverty threshold and the total poverty universe). By 2009, the SAHIE program had matured sufficiently to be considered a production model, and now it produces health insurance coverage estimates on an ongoing basis.

In their paper, O'Hara and Bauder (Chapter 13, in this volume) assess the ACS-based SAHIE model by comparing its state estimates with a CPS-based SAHIE model, and then modifying the model to obtain ACS model-based estimates for more income categories. They observe that gains from model-based state estimates increase as the number of domains increases (e.g., the income categories). These potential changes to the model, based on the strengths of the ACS, are intended to create more useful or refined estimates of uninsured populations for policy makers and other stakeholders.

They also reported on research comparing the ACS-based SAHIE model with the current CPS-based model. The research shows that an ACS-based SAHIE model generally has lower measures of uncertainty compared with a CPS-based SAHIE model. This is not surprising given that 1 year of ACS data is needed rather than a 3-year average of data from the CPS ASEC. The research also tested whether the ACS survey data could support modeling more income-to-poverty ratios. They found that, for most small domains (state/age/race/sex estimates for each income group), the ACS model-based estimates offer a large improvement over using survey-only estimates. With ACS direct estimates, many of the domains at the state level have coefficients of variation that should be used with caution or not at all.

According to the authors, the SAHIE estimates of uninsured people produce promising results, particularly when the ACS estimates are used as input. The SAHIE estimates could be a valuable tool for policy makers to evaluate and administer means-tested programs, such as Medicaid or the Children's Health Insurance Program, and could be used to target funds for outreach to specific uninsured and underserved demographic groups. The state estimates could be used by states to estimate the number of uninsured people by income group, to assess program performance, and to assist in managing the program, for example, to approximate the total cost of subsidizing premiums for the health insurance exchanges.

O'Hara and Bauder report that research on the SAHIE model is continuing. Some planned areas of research include using ACS direct variance estimates. They can produce replicate-based estimates of the variances of ACS estimates, relaxing model assumptions by investigating alternatives to assumptions of independence and of constant variances, developing predictors for insurance coverage, such as those that now predict income status, testing the model to incorporate more income groups (e.g., 250-300 and 300-400 percent of poverty) and age groups (e.g., 19-25-year-olds), and producing multiyear ACS estimates. With these improvements, the possibility becomes more likely of using the results of models to supplement surveys as the official estimates of health insurance coverage for children.

Using Small-Area Models to Understand Survey Differences

In the discussion period following these presentations, Eric Slud, a member of the steering committee, suggested a way in which small-area models could be used in confirming and understanding systematic differences between health-insured estimates from multiple health surveys relating to subpopulations determined by income and age group. His suggestion was that each of the survey estimates of the size of such insured subpopulations could be modeled as a response variable, at levels of aggregation of state and sometimes smaller units, using the other survey estimates (aggregated as necessary) of related quantities as predictors. Some of the survey estimates might turn out to be highly predictable and reproducible from the others, and some might not.

He thought this might be a cost-effective way to document which of the survey estimates are most reproducible and predictable from the others and to calibrate them. Whether this could lead to a combined estimator of still higher quality would be a further worthwhile research issue. He suggested that an exploratory statistical analysis would be much cheaper than survey modifications, giving rise to a hope that this would be an appropriate path of investigation for the Census Bureau.

COMBINING INFORMATION FROM MULTIPLE SURVEYS

Nathaniel Schenker summarized his work on combining information from multiple surveys as a contribution to this session that highlighted alternate ways of strengthening estimates of uncovered children. His paper was based, in large part, on work published in a review article by Schenker and Raghunathan (2007).

He pointed out that there are several reasons for combining information from multiple surveys. They include the possibility of taking advantage of different strengths from different surveys and using one survey to supply information that is lacking from another. There are also reasons associated with handling various types of nonsampling errors, such as coverage error (when the population being sampled is different for some reason from the actual target population), errors due to missing data (nonresponse being one cause of missing data), and measurement or response error (when the variables being measured are measuring them with error).

He reported on three projects that involve combining information: combining estimates from a survey of households (the National Health Interview Survey, NHIS) and a survey of nursing homes (the National Nursing Home Survey, NNHS) to extend coverage; using information from an examination-based health survey (the National Health and Nutrition Examination Survey, NHANES) that uses measurement error models to predict clinical outcomes from self-report answers and covariates in order to improve on analyses of self-reported data in a larger interview-based survey; and combining information from the Behavioral Risk Factor Surveillance System survey (BRFSS) and the NHIS using Bayesian methods in order to enhance small-area estimation.

The combination of data from the NHIS and the NNHS was for the years 1985, 1995, and 1997. The NHIS samples households and collects data with regard to chronic conditions. In the NNHS, nursing homes are sampled and the nursing home staff provide information on the diagnosis of selected residents from medical records. For those aged 85 and over, 21 percent were in nursing homes, so improving coverage of this population required a combination based on design-based prevalence estimates from the two surveys for various chronic conditions.

Schenker's second example used information from the NHANES to improve on analysis of self-reported data, when they might not accurately reflect prevalence of chronic conditions because of the way the questions are asked. The NHANES was used because there is an interview phase and a physical exam, which provides actual clinical measures. In this study, measurement error or imputation models were fitted, and the fitted models were used to impute clinical outcomes for people in the NHIS. This work is akin to developing valid estimates of prevalence for small populations based on interviews in order to improve surveillance without the need for clinical measurements.

The third study used the BRFSS, a large survey with extensive coverage but, as a telephone survey, it does not cover households without phones and has high nonresponse rates. Schenker made the point that this was a form of missing data problem because the data for a part of the population that was not sampled is missing.

These three studies yielded several lessons with possible applicability to the issues related to improving measurement of children's health insurance coverage. First, combining information across surveys can yield gains, particularly when the surveys have complementary strengths. Second, the methods developed can become obsolete quickly, particularly when collection techniques are changing rapidly, as is the case with telephone surveys. Third, attention must be paid to such issues as context and mode, as the phrasing of the questions has a significant impact. And fourth, sample designs count. For example, very different results would be obtained from an area-level design rather than a person-level design.

Some of the more practical lessons learned include the tremendous amount of work that goes into sharing data and estimates among multiple collection agencies, owing to confidentiality concerns and different policies and priorities among the agencies, as well as issues with such matters as selection of software.

DISCUSSION

In the discussion that followed these three presentations on modeling and combining data, the following points were made:

- It is important, as in the SAHIE program, to refit the models every year, since the administrative data could change and affect the model coefficients between periods. This was important in 1997, when the welfare reform legislation changed the Food Stamp Program and affected how the Food Stamp variable worked in the models. The variable was dropped but eventually was reinstated after states settled their administrative procedures.
- There is concern that models may not be an acceptable way of making allotments of CHIP resources, although it was pointed out that Title I education funds are allocated using the results of the SAIPE model and have been for some time.
- It is important when dealing with administrative data in the models to keep in mind the biases that exist among the states that affect the data series. Sometimes these biases are difficult to measure, as they may be related to subtle program management differences.

One such difference is state policies regarding how long children are maintained on the rolls even when no longer actively enrolled, due to 12-month continuous enrollment policies.

- Combining data series, as discussed by Schenker, has other promising attributes. One participant pointed to the possible use of such analysis in assisting states in finding prospective program participants on the basis of the results of the combination of surveys, comprising a more powerful identification tool than is presently possible using only the BRFSS.

7

Looking Ahead

Along with the rest of the nation's health care system, the Children's Health Insurance Program (CHIP) finds itself in a time of massive change. In the 18-month period prior to the workshop, CHIP was reengineered with passage of the Children's Health Insurance Program Reauthorization Act in early 2009; it was then further clarified with passage of the American Recovery and Reinvestment Act (ARRA) in 2009 and the Patient Protection and Affordable Care Act (PPACA) in 2010. During the workshop, participants emphasized that the program can be expected to be in a state of change for some time to come, as changes wrought by the new health care reform legislation are implemented at the national and state levels.

At the same time, on the data front, two major trends are expected to yield a change in the kind of information that will be available to understand the impact of the new health reform legislation on children's health insurance coverage, among other issues. These are the increased attention to the quality of administrative data on the CHIP and Medicaid programs at the federal and state levels and the growing maturity and acceptance of the American Community Survey (ACS). The coming decade is a time in which, with proper planning, there will be new opportunities for developing and implementing improvements to survey and administrative databases in ways that will enhance understanding of children's health insurance coverage issues.

MATCHING DATA

Although this is a time of constrained resources, there are exponentially expanding needs for data that will assist policy makers and program administrators at the federal, state, and local levels ensure coverage of eligible populations, manage programs efficiently, and evaluate the effectiveness of program options and approaches.

One consistent theme of the workshop was the emerging power to link survey to survey, survey to administrative record data, and administrative to administrative record data systems. In addition to hearing about pioneering record matching projects, such as those involving State Health Access Data Assistance Center (SHADAC), the National Center for Health Statistics (NCHS), the Agency for Healthcare Research and Quality (AHRQ), the U.S. Department of Health and Human Services Assistant Secretary for Planning and Evaluation (ASPE), the Centers for Medicare & Medicaid Services (CMS), and the U.S. Census Bureau, the workshop was informed that the CMS has released a task order to produce person-level data files on Medicaid eligibility, service utilization, and payment information to support comparative effectiveness research funded by the ARRA.

As was pointed out in the workshop, as a component of ARRA, CMS has been approved by the U.S. Department of Health and Human Services to produce and enhance the Medicaid Analytic eXtract file (MAX). The agency has released a task order relating to the production of MAX, along with the development of an accelerated version of MAX to produce more timely data for research use. The range of activities that are optional under this task order include verification of Social Security numbers, production of a master file of Medicaid/CHIP enrollment, and linkage of MAX to federal surveys, identified in the solicitation as the National Health Information Survey, the National Health and Nutrition Examination Survey, the Longitudinal Study on Aging, the National Nursing Home Survey, the Medicare Current Beneficiary Survey, and a prototype for the Census Bureau's ACS.

Many of the opportunities for matching exist at the state level. It was suggested that state-level matching could involve using other data to get a better match of the potentially eligible population with enrollment files. These links between administrative data systems would permit improvements in the modeling of state impacts and in the development of measurement error models. Because insurance coverage is dynamic, useful links would be not only cross-sectional but also longitudinal—creating records on individuals over time. One benefit of this matching activity, it was suggested at the workshop, would be to improve understanding of the existing data, their limitations, and how they could be better used for policy analysis.

IMPROVING UNDERSTANDING OF NONFEDERAL AND STATE COVERAGE SOURCES

Improvements to federal and state survey and administrative record systems will go only part of the way toward gaining a better understanding of children's health insurance coverage. To get a complete picture, the coverage represented by private insurers needs to be better understood. It was suggested that there are few links to private insurance enrollment records as well as little knowledge of the effects of private insurance. Some past efforts to develop such links, such as the SNACC project, were unsuccessful in obtaining private insurers' data. States are also looking into using data from all-payer databases in order to have information on both public and private plans, and it was suggested that understanding of the total coverage picture may change when the Internal Revenue Service (IRS) implements its responsibility for ascertaining the health insurance coverage of the population.

The requirement for coverage and the penalty imposed on those who fail to maintain minimum essential health benefits coverage were established by the PPACA, as amended by the Health Care and Education Reconciliation Act of 2010, in new section 5000A of the Internal Revenue Code, and are scheduled to begin in tax year 2014. In general, the legislation requires individuals, beginning in 2014, to maintain health insurance, with some exceptions. Individuals will be required to maintain minimum essential coverage, which includes eligible employer coverage, individual coverage, grandfathered plans, and federal programs, such as Medicare and Medicaid, among others. In order to assess a penalty, IRS may be obtaining information from insurers about an individual's health insurance coverage with an indication of coverage on a monthly basis (the penalties would be applied monthly). This requirement sets up an opportunity for obtaining extensive new information on insurance coverage.

RATIONALIZING THE AMERICAN COMMUNITY SURVEY AND THE CURRENT POPULATION SURVEY

Several factors identified during the workshop appear to drive increased reliance on the ACS as the major survey source of estimates of children's health insurance coverage, although there will continue to be reliance on the Current Population Survey (CPS) or some purposes. This trend will be based, in part, on the fact that the ACS and the CPS results are close for many characteristics and statuses, particularly in terms of family income and uninsurance. The CPS is generally acknowledged to provide better estimates of poverty due to the income questions. That said, the facts that the ACS has better geographic coverage, a larger sample size (and lower standard errors), and the new coverage question mean that it should increasingly become the primary source of the estimates. This has practical consequences, as the workshop was reminded by David Johnson. The Census Bureau now receives about $20 million annually to pay for the additional work associated with collecting the children's health insurance data on the CPS and to fund a program of research and modeling for these data. These funds would be put in jeopardy if the user community shifted to sole use of the ACS as the source of these data.

The discussion at the workshop focused on the meaning of this trend for these major household surveys. A few participants speculated that there would be increasing pressure to improve measurement of other factors other than coverage that affect the uninsured, such as health status and access. Moreover, it was noted, these additional data items would be most useful if they were ready in time to measure the coverage and effectiveness of the post-2014 program changes, although it was recognized that the next window of opportunity for adding or adjusting questions in the ACS will not occur until 2018.

In the near term, the upcoming, near-simultaneous release in September 2010 of the ACS and CPS estimates of children's coverage in 2009 will draw attention to reconciliation and preference issues. Organizations such as the State Health Access Data Assistance Center are focusing on improving understanding of the ACS this year and are engaged in finding out from states what numbers they use so they will be ready for this event.

IMPACT OF THE NEW HEALTH CARE REFORM LEGISLATION

The workshop discussion focused on the critical factor of time in considering necessary changes to data systems. Officials of the U.S. Department of Health and Human Services reminded everyone that many things will change after 2014, including coverage under Medicaid.

The representatives of the Office of the Assistant Secretary for Policy and Evaluation stressed that the federal government must be flexible during this implementation period and that projects should be put in place to perform some timely infrastructure studies as the implementation activities unfold. Such studies might well learn from the infrastructure studies in Massachusetts that were reported at the workshop, which have helped to assess the effects of the health care changes there.

References

Agency for Healthcare Research and Quality. (2009). *MEPS-HC Tables*. Available: http://www.meps.ahrq.gov/mepsweb/data_stats/summ_tables/hc/hlth_insr/2007/alltablesfy.pdf [June 2010].

Agency for Healthcare Research and Quality. (2010). *Medical Expenditure Panel Survey: Survey Background*. Available: http://www.meps.ahrq.gov/mepsweb/about_meps/survey_back.jsp [July 2010].

Baugh, D. (2010). *CMS Programs for Improving the State Administrative Microdata. Medicaid and the Child Health Insurance Program (CHIP).* Presentation at the Workshop on Evaluating Databases for Use in Uninsured Estimates for Children, June 17-18, National Research Council, Washington, DC. Available: http://www7.nationalacademies.org/cnstat/CHIP%20Workshop%20Agenda.pdf [October 2010].

Blewett, L. (2010). *Overview of State Survey Development and Use: What States Need to Monitor Health Insurance Coverage*. Presentation at the Workshop on Evaluating Databases for Use in Uninsured Estimates for Children, June 17-18, National Research Council, Washington, DC. Available: http://www7.nationalacademies.org/cnstat/CHIP%20Workshop%20Agenda.pdf [October 2010].

Blewett, L., and Davern, M. (2006). Meeting the need for state-level health insurance estimates: Use of state and federal survey data. *Health Services Research, 41*(3 Pt 1), 946-975.

Cohen, R.A., Martinez, M.E., and Free, H. (2007). *Insurance Coverage: Early Release of Estimates from the National Health Interview Survey, 2008*. Report from the National Center for Health Statistics. Available: http://www.cdc.gov/nchs/data/nhis/earlyrelease/insur200906.pdf [October 2010].

Davern, M. (2010). *Unstable Ground: Comparing Health Insurance Estimates from National Surveys*. Presentation at the Workshop on Evaluating Databases for Use in Uninsured Estimates for Children. Available: http://www7.nationalacademies.org/cnstat/CHIP%20Workshop%20Agenda.pdf [October 2010].

Davern, M., Rodin, H., Call, K.T., and Blewett, L.A. (2007). Are the CPS uninsurance estimates too high? An examination of imputation. *Health Services Research, 42*(5), 2038-2055.

Davern, M., Klerman, J.A., Baugh, D., Call, K.T., and Greenberg, G. (2009). An examination of the Medicaid undercount in the Current Population Survey (CPS): Preliminary results from record linking. *Health Services Research, 44*(3), 965-987.

Georgetown Center for Children and Families. (2009). *The Children's Health Insurance Program Reauthorization Act of 2009*. Washington, DC: Georgetown Health Policy Institute.

National Center for Health Statistics. (2010). *NHIS Survey Description*. Available: ftp://ftp.cdc.gov/pub/Health_Statistics/NCHS/Dataset_Documentation/NHIS/2009/srvydesc.pdf [July 2010].

National Research Council. (2000). *Small-Area Income and Poverty Estimates: Priorities for 2000 and Beyond*. Panel on Estimates of Poverty for Small Geographic Areas, Committee on National Statistics. Washington, DC: National Academy Press.

Pascale, J., Roemer, M., and Resnick, D. (2009). Medicaid underreporting in the CPS: Results from a record check study. DOI 10.1093/poq/nfp028. *Public Opinion Quarterly, 73*, 497-520.

Rousseau, D., and Kaiser Family Foundation. (2010). *A User's Perspective on Medicaid and CHIP Administrative Data Quality*. Presentation at the Workshop on Evaluating Databases for Use in Uninsured Estimates for Children, June 17-18, National Research Council, Washington, DC. Available: http://www7.nationalacademies.org/cnstat/CHIP%20Workshop%20Agenda.pdf [October 2010].

Schenker, N., and Raghunathan, T.E. (2007). Combining information from multiple surveys to enhance estimation of measures of health. *Statistics in Medicine, 26*, 1802-1811.

Sebelius, K. (2010). *The First Anniversary of the Children's Health Insurance Reauthorization Act*. Available: http://www.hhs.gov/secretary/about/speeches/sp20100204.html [June 2010].

State Health Access Data Assistance Center. (2003). *Yet Another Wild Card in State Budget Deliberations: Federal SCHIP Allocations to States*. Issue Brief 7. University of Minnesota, School of Public Health. Available: http://www.shadac.org/files/IssueBrief7.pdf [August 2010].

Turner, J., Boudreaux, M., and Lynch, V. (2009). *A Preliminary Evaluation of Health Insurance Coverage in the 2008 American Community Survey*. U.S. Census Bureau, Housing and Household Economic Statistics Division. Available: http://www.census.gov/hhes/www/hlthins/data/acs/2008/2008ACS_healthins.pdf [September 2010].

U.S. Department of Health and Human Services. (2010a). *Children's Health Insurance Program Reauthorization Act: One Year Later, Connecting Kids to Coverage*. Available: http://www.insurekidsnow.gov/chip/chipra_anniversary_report.pdf [August 2010].

U.S. Department of Health and Human Services. (2010b). *Connecting Kids to Coverage: Meeting the CHIPRA Challenge*. Request for Task Order Proposal: OS38169. Washington, DC. Available: http://www.hhs.gov/news/press/2010pres/09/20100903a.html [October 2010].

U.S. Department of Health and Human Services, Centers for Medicare and Medicaid Services. (2010). *FY 2010 Online Performance Appendix*. Available: http://oig.hhs.gov/publications/docs/budget/FY2010_online_performance_appendix.pdf [August 2010].

Part II

Background Papers

8

Monitoring Children's Health Insurance Coverage Under CHIPRA Using Federal Surveys

Genevieve Kenney and Victoria Lynch
The Urban Institute

The Children's Health Insurance Program (CHIP) was reauthorized for an additional 4.5 years in February 2009 through the Children's Health Insurance Program Reauthorization Act (CHIPRA, P.L. 111-3). CHIPRA contained a number of provisions designed to expand eligibility for public coverage among children and to increase take-up of coverage among uninsured children who were already eligible for Medicaid and CHIP. This paper assesses possible data sources for monitoring the impacts of CHIPRA on children's health insurance coverage. The following section provides background on CHIP, CHIPRA, and other recent federal policy changes that also have important implications for Medicaid and CHIP coverage for children and their parents. Subsequent sections discuss key research questions and data needs and describe the strengths and limitations of different data sources that are available at the federal level. The closing section suggests ways to improve existing federal surveys so that they provide more useful information for evaluating CHIPRA and related policy changes.

THE CHILDREN'S HEALTH INSURANCE PROGRAM

CHIP was created in 1997 in an effort to close coverage gaps facing low-income families who did not have access to affordable private coverage for their children but had incomes that were too high to qualify for Medicaid. CHIP was funded as a block grant to states but with higher federal matching rates than states typically received under Medicaid. States had flexibility over theIR design of CHIP, including eligibility thresholds, outreach, retention, enrollment policies, and, within parameters set down under the statute, benefits and cost sharing.

All states chose to expand coverage for children through CHIP and implemented policies aimed at simplifying enrollment processes, many of which were also adopted under Medicaid (Kaye et al., 2006). Since the late 1990s, when these policy changes were adopted, uninsured rates have fallen among children, both those made newly eligible for public coverage under CHIP and those who were already eligible for Medicaid (Dubay and Kenney, 2009; Hudson and Selden, 2007). Importantly, gains in health insurance coverage appear to have translated into improvements in access to care and increased preventive care receipt among children (Davidoff et al., 2005; Kenney and Change, 2004; Kenney and Yee, 2007).

Despite this progress, at the time when CHIPRA was passed, research indicated that millions of children were uninsured despite being eligible for Medicaid or CHIP and that many children enrolled in public coverage were not receiving recommended levels of care (Dubay et al., 2007). Moreover, uninsured rates among low-income children continued to vary widely across states (DeNavas et al., 2008).

In an effort to address these gaps, CHIPRA provided states with new tools to address shortfalls in enrollment as well as access and quality. CHIPRA included new outreach and enrollment grants and bonus payments to states that adopted five of eight enrollment/retention strategies and that experienced Medicaid enrollment that exceeded targeted growth rates.[1] States were also given new options to use Express Lane Eligibility options to facilitate eligibility determination and enrollment and for meeting documentation requirements (U.S. Department of Health and Human Services, 2010). CHIPRA allowed states to use federal dollars to cover legal immigrant children who had been in the United States less than 5 years (previously coverage for such children had to be funded exclusively with state funds). It also provided states with additional federal allotments for CHIP to cover the costs of enrolling more eligible children and of expanding eligibility (e.g., to higher income groups). In

[1] These include (1) Adopting 12-month continuous eligibility for all children; (2) eliminating the asset test for children; (3) eliminating in-person interview requirements at application and renewal; (4) using joint applications and supplemental forms and the same application and renewal verification process for the two programs; (5) allowing for administrative or paperless verification at renewal through the use of pre-populated forms or ex parte determinations; (6) exercising the option to use presumptive eligibility when evaluating children's eligibility for coverage; (7) exercising the new option in the law to use Express Lane; and (8) exercising the new options in the law in regard to premium assistance (Georgetown Center for Children and Families, 2009).

addition, CHIPRA included a number of provisions designed to improve access to care and the quality of care for the children served by Medicaid and CHIP.

In addition, both the American Recovery and Reinvestment Act (ARRA) of 2009 and the Patient Protection and Affordability Act (PPACA) of 2010 included provisions that could affect children's coverage.[2] ARRA included enhanced matching rates to states that maintain their Medicaid eligibility thresholds for children and adults, in an effort to induce them to continue supporting Medicaid coverage during the current recession. The enhanced matching rates, which are at least 6.2 percentage points higher than regular matching rates, were implemented on October 1, 2008, extended in August 2010, and are slated to continue with a phase-down period through June 2011.

PPACA contained a number of important policy changes that could affect both Medicaid and CHIP coverage for children. It legislated comprehensive health reform, including an expansion of Medicaid to adults and children up to 133 percent of the federal poverty level (FPL), a maintenance of effort requirement through 2019 on state Medicaid and CHIP coverage for children, the provision of new subsidies for the coverage of families with incomes up to 400 percent of the FPL, the creation of health insurance exchanges, and coverage mandates for both individuals and employers. PPACA also provided 2 additional years of federal funding for CHIP, beyond what was in CHIPRA, through 2015. It is not clear how long states will be able to continue CHIP, given that new federal funds for CHIP are allocated only through 2015 and that the federal matching rates under CHIP rise by as much as 23 percentage points at that time, which means that a given allocation will be spent more quickly.

This paper focuses on the monitoring of children's coverage under CHIPRA. However, as we approach 2014, the policy questions will begin to focus on how the provisions of PPACA affect children's health insurance coverage. Although the questions of interest will expand to reflect the broader, comprehensive reforms adopted under PPACA relative to CHIPRA, the data issues discussed here will still be relevant for assessing PPACA.

MONITORING CHILDREN'S COVERAGE UNDER CHIPRA

As indicated above, CHIPRA gave states new tools, incentives, and resources to expand health insurance coverage to low-income children. Just 1 year after CHIPRA was passed, 15 states had expanded their eligibility thresholds under Medicaid/CHIP, 19 states had removed the ban on

---

[2] Patient Protection and Affordable Care Act of 2010 (P.L. 111-148).

covering legal immigrant children who had been in the country less than 5 years, and 18 states had proposed improvements to their enrollment and retention processes or received approval to use the new Express Lane Eligibility Option. In 2009, 9 states qualified for bonus payments, and $40 million in outreach grants was allocated to 42 states (another $10 million in tribal outreach grants was funded in February 2010). These policy changes are expected to produce changes in the coverage distribution of children made newly eligible for coverage (i.e., immigrant children in states lifting the 5-year ban and children in income groups made newly eligible for coverage in states that expanded eligibility) and children who were already eligible for coverage under Medicaid and CHIP (in states that streamlined eligibility and retention processes, etc.). In order to monitor children's coverage under CHIPRA, the following types of questions regarding children's health insurance coverage and participation in Medicaid and CHIP need to be addressed:

- Did uninsured rates fall among children following enactment of CHIPRA? If so, by how much?
- Did uninsured rates fall more among some groups of children (defined by race/ethnicity, income, age, health status, etc.) than among others? If so, by how much?
- Did uninsured rates fall among Medicaid-eligible children? CHIP-eligible children? Among groups of children made newly eligible for coverage (e.g., immigrant children in the country for less than 5 years, children in income groups targeted by eligibility expansions in particular states, etc.)?
- Did uninsured rates fall more in some states than in others?
- Did Medicaid/CHIP participation rates increase? Were increases greater for some groups of children?
- How much do Medicaid/CHIP participation rates vary across states? Do differences in participation rates across states narrow over time?
- Did Medicaid/CHIP participation rates change over time? Did they increase more in some states than in others?
- How much do Medicaid participation rates differ from CHIP rates? Have those differences narrowed over time?
- Did rates of public coverage increase over time? What was happening to rates of private coverage over the same time period?
- How did rates of public and private coverage change for different groups of children (defined by race/ethnicity, income, age, health status, etc.)?
- How did rates of public and private coverage change across states?

- To what extent are the observed changes in uninsurance, public coverage, and private coverage among children attributable to CHIPRA?
- To what extent can the observed coverage changes be attributed to individual provisions in CHIPRA?
- To what extent can the observed coverage and participation changes be attributed to gains in particular states?
- To what extent can the observed coverage and participation changes be attributable to specific policy changes related to CHIPRA that were adopted?

To address these questions, valid state and national survey estimates of insurance coverage are needed for the period prior to CHIPRA (to establish a baseline) and for several years following the implementation of CHIPRA-related policies but before the major provisions of PPACA are enacted. These estimates can be derived only from survey data because other sources, such as administrative records for Medicaid and CHIP, do not include information about children who are not enrolled. Establishing a pre-CHIPRA baseline is somewhat difficult because, although CHIPRA was enacted in February 2009, a similar reauthorization bill had been voted on several times in 2007; even though it did not ultimately become law until over a year later, states may have begun making eligibility and enrollment changes to their CHIP and Medicaid programs for children in 2007 and 2008 in anticipation that the bill would ultimately pass.

Valid estimates are needed on how uninsured rates and rates of different types of coverage (e.g., employer-sponsored insurance [ESI], private nongroup coverage, Medicaid/CHIP coverage, other coverage) are changing for children age 18 and under. Because of the critical role that states play in designing and implementing their Medicaid and CHIP, it is essential to have precise annual estimates of the distribution of children's coverage in each state.

It is also critical that the survey data that are used to derive valid coverage estimates permit the identification of children who are eligible for Medicaid and CHIP to assess how coverage and participation rates are changing among children who are targeted by those two programs. Without such information, it is not possible to track how well Medicaid and CHIP are doing at reaching eligible children or to assess whether enrollment, retention, and related outreach efforts are increasing participation. Simulating eligibility for Medicaid/CHIP requires information on the child's health insurance unit (such as family size, income, immigration status, etc.) that is used to determine eligibility for Medicaid and CHIP in each state (Dubay and Cook 2009; Kenney et al., 2010b). Finally, assessing the impacts of CHIPRA and related policy changes requires establishing a counterfactual for what would have happened in the absence of a particular policy change or set of policy changes.

FEDERAL SURVEYS FOR MONITORING CHILDREN'S COVERAGE

Four federal household surveys are available to monitor health insurance coverage on an annual basis: the Current Population Survey (CPS), the National Health Interview Survey (NHIS), the American Community Survey (ACS), and the Medical Expenditure Panel Survey (MEPS). In this paper we include some information on the MEPS but focus more on the CPS, the NHIS, and the ACS because they have larger sample sizes that make them better suited for monitoring coverage over time at the state and national levels and for identifying how changes may relate to changes in public policies. The key features of each of these surveys differ with respect to tracking coverage over time nationally, by state, and for types of populations that are of special concern to policy makers.

Three federal agencies are responsible for producing estimates from the CPS, the ACS, the NHIS, and the MEPS. The Census Bureau produces the CPS and the ACS. The National Center for Health Statistics (NCHS), which is part of the Centers for Disease Control and Prevention (CDC), produces the NHIS. The Agency for Healthcare Research and Quality (AHRQ) produces the MEPS. Unfortunately, no federal agency is responsible for publishing tables that include estimates from all four sources or for providing guidance on what to make of the different estimates and which survey to use for different tracking needs. Another problem with the presentation of published estimates is that there is often a lack of clarity about the insurance concept and reference period reflected in the estimate. For example, the official publication of the CPS 2008 coverage estimates includes the calendar year in its table titles but does not note that the uninsured estimate is designed to represent uninsurance throughout the calendar year, or that the Census Bureau advises that the uninsured estimate can be interpreted as representing a point in time—whether that means the interview date (around March of the following year) or some point during the year the survey asks about is not made evident (DeNavas et al., 2009). The official publication of ACS 2008 coverage estimates includes the calendar year in its table titles but one must know that the ACS is a rolling survey to realize that the estimate is for an average day during the calendar year.[3] The official publication of NHIS coverage

[3] U.S. Census Bureau American FactFinder Table B27001. Health Insurance Coverage Status by Age for the Civilian Noninstitutionalized Population. Available: http://factfinder.census.gov/servlet/DatasetTableListServlet?_ds_name=ACS_2008_1YR_G00_&_type=table&_program=ACS&_lang=en&_ts=294308138878 [October 2010].

estimates indicate which uninsurance estimates cover which time periods ("time of the interview," "at least part of the past year," and "more than a year"), but many people monitoring coverage may not know that these represent estimates of an average day during the calendar year (Cohen and Martinez, 2009). The table titles for the MEPS coverage estimates available on the AHRQ website indicate the time period ("first half of" calendar year or simply the calendar year) but do not indicate that the estimates represent uninsurance throughout those periods. This is especially problematic for anyone just looking at one of the several calendar years for which only half-year estimates are available, because it is less likely that the data user will figure out (i.e., by comparing calendar year estimates) that the half-year estimates are not point-in-time and thus not comparable to estimates from most other surveys.[4]

There is no consensus on how many children are uninsured at a point in time or throughout the year (Office of the Assistant Secretary of Planning and Evaluation, 2003). For example, the most recent year for which full-year estimates of uninsurance are available for more than one survey (assuming that the CPS estimate is not a valid measure of full-year uninsurance) is 2007, and those show a range from 3.7 million in the NHIS to 7.9 million in the MEPS (Cohen et al., 2007).[5] Not only is there disagreement about how many children lack health insurance coverage at a particular point in time nationally, but state-level estimates vary across surveys as well (Blewett and Davern, 2006; Call et al., 2007).

In terms of coverage estimates, the main methodological differences between the CPS, the NHIS, and the ACS relate to sample size, level of detail used in questions collecting information about coverage, other subjects asked about, characteristics of the interview, and postcollection processing (Davern et al., 2009). Other features of sampling and the particular population controls used in weighting may also explain differences in the survey estimates and their suitability for monitoring coverage. We discuss the validity of the NHIS, the CPS, and the ACS in terms of likely misclassification of coverage type, particularly Medicaid/CHIP, given current research findings and ongoing questions about the validity of coverage estimates. We also discuss sampling design, the questions included, and the validity of variables used to study key population subgroups of interest, such as children who are eligible for Medicaid/CHIP.

[4] Online MEPS-HC tables available: http://www.meps.ahrq.gov/mepsweb/data_stats/quick_tables_results.jsp?component=1&subcomponent=0&year=2007&tableSeries=4&searchText=&searchMethod=1&Action=Search [June 2010].

[5] Online MEPS-HC tables available: http://www.meps.ahrq.gov/mepsweb/data_stats/summ_tables/hc/hlth_insr/2007/alltablesfy.pdf [June 2010].

National Health Interview Survey

There is a general consensus that the NHIS produces the most valid coverage estimates (Kenney, Holahan, and Nichols, 2006).[6] The NHIS is a health-focused survey that includes many features to aid respondents in understanding the coverage question and recalling details required to correctly answer it. The NHIS features that may strengthen validity include

- area sample frame;
- well-trained interview staff that work exclusively on this survey;
- fairly high response rate;
- usually an in-person interview;
- a knowledgeable respondent (interviewers encourage older children to report about themselves but indicate that they want to speak about coverage with individuals who are knowledgeable about the coverage status of household members);
- a questionnaire that defines concepts and probes respondent memory as it collects information;
- breadth of content on other health-related data, which potentially helps respondents understand distinctions between coverage types and accurately classify the coverage status of the individuals whom they report about;
- asking about coverage source at the time of the survey, which is associated with lower measurement error;
- asking about Medicaid and CHIP using state-specific names;
- a low level of item nonresponse on insurance sequence;
- asking for many details about coverage (e.g., type of managed care, copayments, deductibles, need for referrals), which may help define relevant concepts and help respondent recall coverage details;
- asking about periods without coverage and when the child last had coverage (for use in estimating full-year uninsurance) and why it stopped (potentially helping the respondent to recall more details required to determine the child's true coverage status);
- verifying no Medicaid for children with no reported coverage;
- asking about citizenship, place of birth, and family relationship, which are some of the important variables needed to simulate eligibility in Medicaid and CHIP;
- asking about medical visits and other uses of coverage or evidence of acting uninsured;

---

[6] Results from the Medicaid Undercount project suggest that underreporting of Medicaid/CHIP is lower in NHIS than CPS; available: http://www.census.gov/did/www/snacc/ [October 2010].

- asking for the name of the insurance plan so the name can be matched to a list of insurance plans by state in a postcollection data processing phase and potentially used to recode misreported coverage type; and
- has been in production for many years and with attention to maintaining a credible time series.

There are also important limitations for using the NHIS to monitor coverage. The most problematic of these is the sampling design, which limits the geographic and other subpopulation estimates that are possible as well as raising validity questions. First, the sample size is too small to produce precise annual state (and substate) estimates for most states. Second, most states have only a very small number of primary sampling units, a fact that raises concerns about the representativeness of the state-level estimates produced by the survey.[7] Third, because of data confidentiality concerns, access to state identifiers is available only through data centers. The ability to use the NHIS to simulate Medicaid/CHIP eligibility is also limited by the quality of the income data, as well as the possible underreporting of the Medicaid/CHIP information coverage, despite all the efforts to measure coverage accurately (U.S. Census Bureau, 2009). The NHIS is also limiting because of the timing of the data release and what is excluded from the published estimates. There is an early release that enables some important coverage evaluations before the survey is fully prepared; however, it is still about 9 months after the interviews are completed, it does not include published estimates for children aged 0-18 separately, and it does not provide valid estimates by income.

Current Population Survey

The CPS has historically played an important role in monitoring coverage. Besides being a relatively large survey using high-quality data collection methods, it is an income- and employment-focused survey and is considered to have valid data on those domains, which are integral for eligibility simulations and other coverage-related analyses. Features that may strengthen validity include

- area frame;
- well-trained interview staff working exclusively on this survey;
- telephone and in-person interviews;
- high response rates;

[7] NHIS PSUs cover only about 25 percent of U.S. counties.

- sample size is large enough for precise state estimates for large states annually;
- family-level questioning about coverage source, which helps get more coverage reported in large households;
- state-specific names and separate questions about Medicaid and CHIP;
- other probes and definitions (the Civilian Health and Medical Program of the Department of Veterans Affairs [CHAMPVA], direct purchase);
- a question on directly purchased coverage that emphasizes that it is not related to a current or former employer;
- asking for detailed information about coverage, including who is the policy holder, who is covered by the same policy, who is covered by someone outside the household, and employer contributions;
- asking several times about any other type of coverage not yet talked about;
- verifying the absence of insurance coverage;
- logical coverage edits performed by the Census Bureau to correct some likely reporting errors;
- asking about citizenship, place of birth, family relationship, supports from people outside the household, firm size, as well as income and employment-related factors, which are some of the important variables needed to simulate eligibility in Medicaid and CHIP;
- asking about health status;
- has been in production for many years and with attention to maintaining a credible time series; and
- the release of estimates and public-use files with state identifiers 5 to 6 months after the data are collected.

The most critical limitation of using the CPS to monitor coverage is the known measurement error with the coverage questions because of confusion, recall bias, and other issues with the retrospective reference period (Pascale, Roemer, and Resnick, 2009). As a result of these, there is more apparent underreporting of Medicaid/CHIP coverage and considerable uncertainty about what the estimates mean, especially compared with other surveys (Davern et al., 2009a; Kincheloe, Brown, and Frates, 2006). In addition, the sample size and the number of primary sampling units is small in many states, which raises concerns about the representativeness and precision of the state estimates. Historically there has also been concern about bias in the imputation process for coverage variables; however, new imputations are being implemented at the Census Bureau to address this problem (Davern et al., 2007). The CPS is also missing information about access to, the need for, and use of health services and spending. The 9-month lag between the end of the calendar year reference period and the release of the published estimates and edited data limits the ability to track coverage in real time; however, the estimates and data are more timely if they are interpreted as representing some time closer to the interview data in March (just 6 months before the release). The published estimates are also limiting because they do not include children aged 0-18.

American Community Survey

The ACS, an annual survey designed to provide intercensal estimates of the information contained on the decennial census long form, added information on health insurance coverage in 2008. Although the ACS is still too new of a resource for studying trends in children's health insurance coverage, it has a number of important strengths relative to the other surveys:

- The most important strengths of the ACS are its very large sample and its sample frame (which samples every county and census tract in the country), allowing for:
  - 1-year coverage estimates for areas with a population of 65,000 or more;
  - starting in 2011, 3-year coverage estimates for areas with populations of 20,000 or more; and
  - starting in 2013, 5-year coverage estimates for all statistical, legal, and administrative entities.
- The coverage information refers to the time of the survey.
- It is possible to put together a variety of substate estimates, including public-use microdata areas, large counties, large metropolitan areas, etc.
- Comparisons with the employment-focused CPS suggest that the ACS also has fairly robust income- and employment-related data, for use in eligibility simulations and other studies.
- Although most data are collected by mail, the Census Bureau computes a 98 percent response rate, which is very high (Griffin and Hughes, 2010).
- The release of estimates and public-use files with state identifiers (8 to 9 months after the end of the survey period, implying an average lag between data collection and data release of 14 to 15 months).

Since the ACS coverage data are new as of 2008, the survey cannot provide an extended pre-period for studying trends in children's coverage prior to the adoption of CHIPRA-related policy changes. Moreover, research is just now being conducted on the validity of the ACS estimates. Although overall the unadjusted ACS estimates of uninsured children were close to the CPS estimates, they were somewhat higher than the NHIS estimates for the same period, and reports of direct purchase of insurance on the ACS are very high (Turner, Boudreaux, and Lynch, 2009). A major limitation of the ACS for monitoring coverage is that much of the data is collected by mail (56 percent of responses in 2008), which means that most respondents complete the survey without the aid of an interviewer. Another major concern is that the coverage question includes no distinction among Medicaid, CHIP, and other sources of government insurance, and no state-specific names for Medicaid/CHIP were provided in 2008 (they were available to interviewers in computer-assisted modes starting in 2009). In addition, there is only one itemized list of coverage types (rather than a detailed series of patterned questioning, defining, and probing, as in the NHIS and the CPS), which could also introduce more measurement error in the reporting of coverage type. Also of concern is the absence of a statement that insurance purchased directly should not have anything to do with a current or former employer as well as the absence of questions about coverage details (managed care, premiums, employer contributions, and other questions that probe memory, define concepts, and can be used to recode misreports). In addition, the ACS does not include a verification of uninsurance or questions about duration of uninsurance. Another concern with the 2008 ACS estimates is that there was relatively little postcollection processing on the ACS to remedy possible reporting error. By contrast, the NHIS, for example, gives field representatives the opportunity to indicate concern about the validity of coverage reports and also collects the name of the person's plan and uses it to reclassify coverage type. The CPS, for another contrasting example, uses other coverage-related information collected about the person or family to reclassify coverage on a logical basis.

The ACS has a number of other content limitations. In particular, family relationship information is not directly available for analysis, making it much more difficult to identify health insurance units for eligibility simulations; there is no information on the child's general health status or the parents' firm size. And while the ACS sample is very large and its published estimates cover a variety of important geographic areas, the sample released for public-use data is smaller and excludes many geographic identifiers (e.g., congressional districts), which makes it more difficult to track meaningful coverage changes for smaller states and smaller subgroups, short of gaining access to a Census Bureau data center (which requires a comprehensive application that takes several months and must meet stringent requirements).

IMPROVING FEDERAL SURVEYS

Published estimates and public-use files could be made more useful for monitoring children's coverage in a number of ways:

- Modify questionnaires to address known problems with coverage validity and collect other information needed to monitor coverage, especially as coverage options change with reforms:
  - CPS: Add a question about current coverage (being tested now);
  - NHIS: Expand income series (some changes are currently under way);
  - NHIS and MEPS: Include a variable for the coverage status shown on the insurance card that respondents are asked to collect at the beginning of the interview; and
  - ACS: Add a clarification to the itemized question about direct purchase (to emphasize that it is coverage that is unrelated to a current or former employer); add state-specific Medicaid/CHIP names (if adding the actual name is not feasible, add "CHIP or the children's health insurance program in your state" or add an insert that includes a list of the Medicaid/CHIP names in different states); include more definitions of coverage types in the booklet of directions for mail respondents/interviewers and refer to their availability in the introduction to the health insurance question; add questions on firm size and general health status, verification of uninsurance, any government assistance paying for health insurance premiums (primarily for use in recoding coverage), and health insurance plan name (also primarily for use in recoding coverage).
- Perform call-backs of selected cases in the ACS and the CPS reported to have direct purchase, starting with those identified as low-income or logically covered by Medicaid, military, or Employee State Insurance (ESI). Use results to edit erroneous reports and to refine rules for logical coverage editing.
- Use all the explicit and implied information about coverage that is collected about each child and his or her family to create an edited set of coverage variables if other, more reliable reported information implies the original coverage variable is incorrect (Lynch, Boudreaux, and Davern, 2010).
  - ACS and CPS: Create an edited version of the variable for directly purchased coverage that excludes sample people who appear to have coverage from Medicaid/CHIP or the military or other employer. Extend current logical editing rules for Medicaid/CHIP. Our research indicates that such rules for the 2008 ACS

reduces the uninsured estimate for children aged 0-18 from 8.2 to 7.3 million (which is very close to the NHIS uninsured estimate) and increases the estimated number of children with Medicaid/CHIP as their primary coverage on the ACS from 19.8 to 24.4 million, which is 6.0 percent lower than the comparable administrative count for June 2008 (Lynch, 2010).

- Conduct more research on administrative records to identify reasons for errors in reporting children's coverage and provide data users with methods to adjust for them:
  - o ACS, CPS, NHIS, and MEPS: Adopt methods from the Medicaid Undercount project's research on reporting errors about enrollees of all ages (in the CPS, the NHIS, and MEPS) to research on just children (U.S. Census Bureau, 2008); and
  - o ACS, CPS, NHIS, and MEPS: Develop models to adjust children's coverage estimates and make them available to data users, as has been done for all ages in the CPS and the NHIS (Davern et al., 2009b). Test validity of logical coverage edits against administrative data.
- Conduct targeted methodological research to identify survey features that can be modified to reduce reporting errors about coverage:
  - o ACS: Reassess reports of direct purchase and the method of assigning responses that are written in or reported as other coverage. For example, do not code write-in cases with reported/logical ESI, military, or Medicaid/CHIP as also having direct purchase. Assess how Massachusetts sample children with subsidized and unsubsidized coverage through the Health Connector are being reported and use findings to refine data collection strategies aimed at identifying children who end up obtaining coverage through health insurance exchanges under reform. Reexamine terms used to describe coverage types in mail mode and assess respondent ability to correctly classify coverage; and
  - o CPS, NHIS, and MEPS: Examine causal mechanisms for factors identified as predictors of Medicaid misreport.
- Conduct more interagency research on differences in coverage and other variables needed to study coverage:
  - o Explain differences in estimates of coverage (especially nongroup coverage and Medicaid/CHIP); and
  - o Explain differences in estimates of family income from the ACS, the CPS, and the NHIS.
- Present published estimates that cross-reference the other federal estimates, explain possible reasons for discrepancies across the

surveys, cover more policy-relevant constructs, and include more explicit and accurate labels:

- ACS: Re-release 2008 estimates with the logical coverage edits adopted for 2009 and beyond. Describe published estimates as point in time or an average day in the calendar year;
- MEPS: Improve the policy relevance of published estimates: add point-in-time, part-year, and full-year estimates for those recent years that do not publish those estimates; link to the NHIS (the MEPS sampling frame) to provide more information on changes in coverage;
- CPS: Explicitly describe published estimates as approximately point in time or an average across the prior year; and
- ACS, CPS, NHIS, and MEPS: Include estimates for the different policy-relevant definitions of "children," meaning both for children aged 0-18 and 0-17. Release published estimates with an introduction that informs policy makers and other users about complexities, including the likely possibility that measuring coverage is becoming more complex as the types and numbers of plans increase; the fact that the time frame for a person's coverage status is important because a person's health insurance status can change over time; the fact that how coverage type is defined is important because individuals may categorize their status differently from technical definitions; and the fact that these complexities are part of the reason estimates differ by survey.

- Include more documentation about measurement problems and provide more information in published materials to help readers correctly interpret and understand estimates.
- Give data users more information so they can estimate more concepts and have more flexibility dealing with limitations and comparing across surveys:
  - ACS: provide the month of interview and person-level rural/urban or metropolitan variables (using as up-to-date information on metropolitan statistical area boundaries or local area population density as possible);
  - CPS and ACS: Provide flags for logical editing that include an indicator of the reason for an edit; and
  - NHIS: Provide flags for reason for coverage recode.

It is important to recognize that although some of these changes are geared toward improving the validity of the coverage estimates for children, they could also introduce breaks in the time series for a particular survey. Thus, in order to assess changes in coverage over time, it would be important to make needed adjustments to the estimates so that they are comparable before and after the changes are made. In addition, we emphasize recommendations that are feasible to implement without requiring a large increase in funding. However, if we broadened the scope to include important policy questions related to access to care and service use among children, we would strongly recommend increasing the capacity of the federal surveys (such as the NHIS) to produce valid state estimates on these questions. This would require both changing the survey's sample frame or including more primary sampling units and increasing the number of children included in the sample each year. Such an expansion would provide important information about how well individual states are doing at achieving the ultimate objective of CHIPRA and of health reform more broadly, which is to improve the health and functioning of children and adults. However, it would be critical for the state identifiers to be released so that states' progress could be studied with public-use files.

We have focused on measures designed to improve the information available on insurance coverage for children in the current coverage environment. However, it will be important for federal surveys to anticipate the new coverage options that will be available under health reform and that they adjust survey questions and content accordingly to allow the tracking of coverage at the national and state levels and for key population subgroups. Given that the major pieces of health reform are not slated to be implemented until 2014, there is time to test out new questions and to coordinate questionnaire changes across surveys so that they are in place in 2012-2013.

ACKNOWLEDGMENT

We appreciate the helpful feedback from the other participants at the Workshop on Evaluating Databases for Use in Uninsured Estimates for Children. This paper reflects the views of the authors and does not necessarily represent the views of the Urban Institute, its sponsors, or its trustees.

REFERENCES

Blewett, L.A., and Davern, M. (2006). Meeting the need for state-level health insurance estimates: Use of state and federal survey data. *Health Services Research, 41*(3 Pt 1), 946-975.

Call, K.T., Davern, M., and Blewett, L.A. (2007). Estimates of health insurance coverage: Comparing state surveys with the Current Population Survey. *Health Affairs, 26*(1), 269-278.

Cohen, R.A., and Martinez, M.E. (2009). *Health Insurance Coverage: Early Release of Estimates from the National Health Interview Survey, 2008.* Report from the National Center for Health Statistics. Available: http://www.cdc.gov/nchs/data/nhis/earlyrelease/insur200906.pdf [October 2010].

Cohen, R.A., Martinez, M.E., and Free, H. (2007). *Health Insurance Coverage: Early Release of Estimates from the National Health Interview Survey, 2008.* Report from the National Center for Health Statistics Available: http://www.cdc.gov/nchs/data/nhis/early release/insur200806.pdf [October 2010].

Davern, M., Rodin, H., Call, K.T., and Blewett, L.A. (2007). Are the CPS uninsurance estimates too high? An examination of imputation. *Health Services Research, 42*(5), 2038-2055.

Davern, M., Quinn, B., Kenney, G., and Blewett, L.A. (2009a). The American Community Survey and health insurance coverage estimates: Possibilities and challenges for health policy researchers. *Health Services Research, 44*(2), 593-605.

Davern, M., Klerman, J.A., Baugh, D., Call, K.T., and Greenberg, G. (2009b). An examination of the Medicaid undercount in the Current Population Survey (CPS): Preliminary results from record linking. *Health Services Research, 44*(3), 965-987.

Davern, M., Klerman, J.A., Ziegenfuss, J., Lynch, V., Baugh, D., and Greenberg, G. (2009c). A partially corrected estimate of Medicaid enrollment and uninsurance: Results from an imputational model developed off linked survey and administrative data. *Journal of Social and Economic Measurement, 34*(4), 219-240.

Davidoff, A., Kenney, G., and Dubay, L. (2005). Effects of the State Children's Health Insurance Program expansions on children with chronic health conditions. *Pediatrics, 116*(1), e34-e42.

DeNavas, C., Proctor, B.D., and Smith, J. (2009). *Income, Poverty, and Health Insurance Coverage in the United States: 2008.* U.S. Census Bureau. Current Population Reports, series P60-229. Washington, DC: U.S. Government Printing Office.

Dubay, L., and Cook, A. (2009). *How Will the Uninsured Be Affected by Health Reform*? Washington, DC: Kaiser Commission on Medicaid and the Uninsured.

Dubay, L., and Kenney, G. (2009). The impact of CHIP on children's health insurance coverage: An analysis using the National Survey of America's Families. *Health Services Research, 44*(6), 2040-2059.

Dubay, L., Guyer, J., Mann, C., and Odeh, M. (2007). Medicaid at the ten-year anniversary of SCHIP: Looking back and moving forward. *Health Affairs, 26*(2), 370-381.

Georgetown Center for Children and Families. (2009). *The Children's Health Insurance Program Reauthorization Act of 2009.* Washington, DC: Georgetown Health Policy Institute.

Griffin, D., and Hughes, T. (2010, May). *Mixed Mode Data Collection in the American Community Survey.* In the American Association of Public Opinion Research Conference Proceedings. Chicago, IL.

Hudson, J., and Selden, T. (2007). Children's eligibility and coverage: Recent trends and a look ahead. *Health Affairs, 26*(5), w618-w629.

Kaye, N., Pernice, C., and Cullen, A. (2006). *Charting SCHIP III: An Analysis of the Third Comprehensive Survey of State Children's Health Insurance Programs.* Portland, ME: National Academy for State Health Policy.

Kenney, G., and Change, D. (2004). The State Children's Health Insurance Program: Successes, shortcomings, and challenges. *Health Affairs, 23*(5), 51-62.

Kenney, G., and Yee, J. (2007). SCHIP at a crossroads: Experiences to date and challenges ahead. *Health Affairs, 26*(2), 356-369.

Kenney, G., Holahan, J., and Nichols, L. (2006). Toward a more reliable federal survey for tracking health insurance coverage and access. *Health Services Research, 41*(3), 918-945.

Kenney, G., Lynch, V., and Cook, A. (2010, June). *Medicaid and CHIP Eligibility and Participation rates.* Urban Institute, ACS working paper.

Kenney, G., Lynch, V., Cook, A., and Phong, S. (2010). Who and where are the children yet to enroll in Medicaid and the Children's Health Insurance Program? *Health Affairs, 29*(10), 1920. Available: http://content.healthaffairs.org/cgi/reprint/29/10/1920 [October 2010].

Kincheloe, J., Brown, E.R., and Frates, J. (2006). Can we trust population surveys to count Medicaid enrollees and the uninsured? *Health Affairs, 25*(4), 1163-1167.

Lynch, V. (2010). *Applying Logical Coverage Edits for Analyzing Medicaid/CHIP Participation and Coverage in the 2008 ACS, Revised Edition*. Washington, DC: Health Policy Center.

Lynch, V., Boudreaux, M., and Davern, M. (2010, July). *Applying and Evaluating Logical Coverage Edits to Health Insurance Coverage in the American Community Survey*. Census Bureau, Housing and Household Economic Statistics Division working paper.

Office of the Assistant Secretary for Planning and Evaluation (ASPE). (2003). *Understanding Different Estimates of Uninsured Children: Putting the Differences in Context*. U.S. Department of Health and Human Services working paper, May 2003.

Pascale, J., Roemer, M., and Resnick, D. (2009). Medicaid underreporting in the CPS: Results from a record check study. *Public Opinion Quarterly, 73*, 497-520.

Turner, J., Boudreaux, M., and Lynch, V. (2009). *A Preliminary Evaluation of Health Insurance Coverage in the 2008 American Community Survey*. U.S. Census Bureau, Housing and Household Economic Statistics Division working paper, October 2009.

U.S. Census Bureau. (2008). *SNACC Research Project to Understand the Medicaid Undercount. The University of Minnesota's State Health Access Center, the Centers for Medicare & Medicaid Services, the Department of Health and Human Services Assistant Secretary for Planning and Evaluation, the National Center for Health Statistics, and the U.S Census Bureau. Phase II Research Results: Examining Discrepancies Between the MSIS and the CPS ASEC*. Available: www.census.gov/did/www/snacc/SNACC_Phase_2_report.pdf [October 2010].

U.S. Census Bureau. (2009). *SNACC Research Project to Understand the Medicaid Undercount. The University of Minnesota's State Health Access Center, the Centers for Medicare & Medicaid Services, the Department of Health and Human Services Assistant Secretary for Planning and Evaluation, the National Center for Health Statistics, and the U.S Census Bureau. Phase IV Research Results: Estimating the Medicaid Undercount in the National Health Interview Survey (NHIS) and Comparing False-Negative Medicaid Reporting in NHIS to the Current Population Survey (CPS)*. Available: http://www.census.gov/did/www/snacc/docs/SNACC_Phase_IV_Full_Report.pdf [October 2010].

U.S. Department of Health and Human Services. (2010). *Children's Health Insurance Program Reauthorization Act: One Year Later, Connecting Kids to Coverage*. Available: http://www.insurekidsnow.gov/chip/chipra_anniversary_report.pdf [October 2010].

9

Health Insurance Coverage in the American Community Survey: A Comparison to Two Other Federal Surveys

Joanna Turner and Michel Boudreaux
University of Minnesota, State Health Access Data Assistance Center

With the passage of national health care reform legislation, there is a growing need for in-depth information on health insurance coverage across key population subgroups and across state and local areas. The American Community Survey (ACS), which added a question about health insurance coverage in 2008, has the potential to serve as a major information resource to support the implementation and evaluation of health care reform at the federal, state, and local levels.

This paper describes the ACS health insurance data, discusses some of the methodological issues that arise in collecting them, and shows key estimates from the ACS compared with estimates from other federal surveys, paying particular attention to estimates for children. In comparing the ACS estimates with other data sources, we also pay particular attention to the impacts of differences in question design and data processing of the estimates.

The paper begins with an overview of the ACS, including a discussion of the health insurance question. The next section compares key uninsurance estimates from the ACS with estimates from the Annual Social and Economic Supplement to the Current Population Survey (CPS ASEC) and the National Health Interview Survey (NHIS), briefly discussing challenges in collecting health insurance coverage in the surveys and the likely implications of differences in data processing. The final section summarizes the implications of our findings for the role of the ACS in tracking insurance coverage over time. This paper recaps the preliminary analysis of Turner, Boudreaux, and Lynch (2009), building on that work to further explore differences among the surveys. The focus here is on providing an overview for analysts who may not be familiar with the ACS, the CPS ASEC, or the NHIS estimates of health insurance coverage.

AMERICAN COMMUNITY SURVEY

The ACS is a nationwide survey designed to collect and produce economic, social, demographic, and housing estimates on an annual basis. It is conducted in all U.S. counties and Puerto Rico *municipios* and participation is required by law. About 3 million housing unit addresses are sampled annually throughout the United States and Puerto Rico. There are separate housing unit (HU) and group quarters (GQ) samples. GQ include nursing homes, correctional facilities, military barracks, and college/university housing, among others.

The sample coverage of the ACS is different from other surveys that gather information about health insurance coverage. For example, neither the CPS ASEC nor the NHIS samples institutional GQ, residents of Puerto Rico, or active-duty military members. Furthermore, neither the CPS ASEC nor the NHIS uses a sample that draws from every county equivalent across the United States.

ACS data are collected continuously using independent monthly samples. The ACS uses three modes of data collection for HUs. All sampled households are mailed a paper survey. All mail nonrespondents are followed up with an attempt for a telephone interview, and a sample of roughly one in three telephone nonrespondents is followed up with personal visits. The nonresponse interviews are conducted using computer-assisted instruments—computer-assisted telephone interviewing (CATI) or computer-assisted personal interviewing (CAPI).[1]

Respondents living in GQ facilities complete their forms using a different procedure based on the size of the facility. Some respondents fill out the paper form, and some forms are completed by field representatives.

The application of sequential modes in tandem with mandated responses leads to a high unit response rate. The official response rate is derived by estimating the number of completed interviews over the estimated number of units that should have been interviewed. This procedure omits units that are ineligible for the survey (such as businesses) and those that did not respond to the mail or phone interview but were not sampled for personal visit follow-up. In 2008, the response rate was 98 percent.[2] This approach to calculating the response rate has appealing

[1] Telephone nonresponse follow-up is conducted through CATI, and personal visit nonresponse follow-up is conducted through CAPI.

[2] Available: http://www.census.gov/acs/www/acs-php/quality_measures_response_2008.php [May 2010].

qualities when using the response rate as a proxy for nonresponse bias. However, alternative measures of the response rate may also be helpful. The Census Bureau publishes the number of originally sampled units and the number of final interviews for each state. For the United States as a whole, 2.8 million HU addresses were selected, and 1.9 million addresses completed an interview.[3]

The Census Bureau publishes ACS single-year estimates for areas with populations of 65,000 or more, 3-year estimates for areas with populations of 20,000 or more, and 5-year estimates for all statistical, legal, and administrative entities. The health insurance coverage data and all new content added to the 2008 questionnaire will have the first 3-year estimates released in 2011, based on 2008-2010. The first release of 5-year estimates will be in 2013, based on 2008-2012.

Multiyear estimates from the ACS differ from average annual estimates from the CPS ASEC. The multiyear data from ACS represent a period estimate derived from data collected for 3- and 5-year periods.[4] The CPS ASEC multiyear estimates are 2- and 3-year averages, but they are not based on pooled data. Thus, while the ACS and the CPS employ a conceptually similar moving-average concept, they are distinct.[5]

Health Insurance Coverage Question

The ACS questionnaire has two sections. In the housing characteristics section, the respondent answers questions for the household. In the personal characteristics section, the respondent answers a set of person-level questions for each member of the household. The health insurance coverage question is asked for each person in the household. The respondent is instructed to report each person's current coverage by marking "yes" or "no" for each of the eight coverage types listed (labeled as subparts a to h). The question text is reproduced below:

> Is this person CURRENTLY covered by any of the following types of health insurance or health coverage plans? Mark "Yes" or "No" for EACH type of coverage in items a–h.

---

[3] Retrieved on May 13, 2010, from: http://www.census.gov/acs/www/acs-php/quality_measures_sample_2008.php [July 2010].

[4] More detail on the interpretation of ACS estimates is available in "Statistical Issues of Interpretation of the American Community Survey's One-, Three-, and Five-Year Period Estimates" at: http://acsweb2.acs.census.gov/acs/www/Downloads/MYE_Guidelines.pdf [July 2010].

[5] Available at: http://www.census.gov/acs/www/Downloads/JSM2007_Beaghen_Weidman.pdf [July 2010].

a. Insurance through a current or former employer or union (of this person or another family member)
b. Insurance purchased directly from an insurance company (by this person or another family member)
c. Medicare, for people 65 and older, or people with certain disabilities
d. Medicaid, Medical Assistance, or any kind of government-assistance plan for those with low incomes or a disability
e. TRICARE or other military health care
f. VA (including those who have ever used or enrolled for VA health care)
g. Indian Health Service
h. Any other type of health insurance or health coverage plan.

The respondent is asked to write in the type of coverage for household members reported to have another type of health insurance or health coverage plan in item h. The health insurance question is intended to capture comprehensive plans.[6] Plans that cover only specific health services, such as dental plans, or are limited to coverage due to an accident or disability are not considered health insurance coverage. Furthermore, it is important to note that subpart d intends to capture all public health insurance programs and is not just an estimate of Medicaid coverage.

Missing responses to the question subparts a to g were assigned a "yes" or "no" response through editing and hot-deck imputation. During the editing process, write-in answers describing or naming the type of "other" health insurance or health coverage plan in subpart h were classified into one of the first seven categories. Hence, only the first seven types of health coverage are part of the microdata file; subpart h and the write-in are not included.

Using the complete edited data, people were considered insured if they had a "yes" in at least one of the coverage types: employer- or union-based plan; a private plan purchased directly; military health care; Medicare, Medicaid, or other public programs; or veterans (VA) health care. People who had no reported health coverage or whose only health coverage was Indian Health Service are considered uninsured. Indian Health Service alone is not considered comprehensive coverage (State Health Access Data Assistance Center, 2005). The types of health insurance are not mutually exclusive; people may be covered by more than one type at the same time.

[6] A guide to help the respondent complete the survey form is provided along with the paper questionnaire.

Item Completeness and Imputation Rates

This section examines patterns of nonresponse and imputation rates for the 2008 ACS. Respondents with complete item response had a "yes" or "no" to each of the first seven types of coverage (subparts a through g on the mail questionnaire). Respondents with no complete items had neither a "yes" nor a "no" to all seven items. The remainder had partial response, meaning that the respondent had at least one item with "yes" or "no," but not all. This analysis of nonresponse does not include write-in responses and excludes respondents that were sampled in 2007 and returned their paper survey in 2008.[7] The percentage of people with responses (either "yes" or "no") to all, some, and none of the seven item subparts is presented in Table 9-1. Across all modes, 73.0 percent of people had a "yes" or "no" response to each item; 23.2 percent responded to at least one but not all items, and 3.8 percent left all the items blank.

This varied by mode, with mail respondents the least likely to provide complete item response, at 51.8 percent. In the GQ population, 81.0 percent had complete health insurance data. People in HUs interviewed by telephone (CATI) or in person (CAPI) were the most likely to give complete item response, at 96.1 percent. This pattern reflects both differences in the instruments and differences in the composition of people in each mode.

In addition to classifying the write-in responses, the editing process applied logical edit rules. If a respondent marked "yes" to one and only one of the types and all other subparts were left blank, the types associated with the blanks were assigned values of "no." For example, a respondent marked "yes" for employer-provided coverage (subpart a) and left the rest blank. The edited final response for that person would be a "yes" for employer- or union-based coverage and a "no" to all of the others: direct purchase, Medicare, Medicaid, military health care, VA, and Indian Health Service. The assumption was made that if a respondent checked one of the types of coverage as "yes" and left the rest blank, that these blanks were implied "no's." This process turned some partial responses into complete responses, and they were not considered imputed. This editing choice was the result of analysis of the pattern of responses to the paper form. Table 9-1 also presents the weighted allocation rate—the percentage of people who had an answer to at least one of the health insurance types obtained through hot-deck imputation. In the population considered, 9.7 percent had at least one health insurance variable imputed.

[7] Just over 27,000 forms, although included in the 2008 data, actually used the 2007 instrument. Thus, they could not answer the health insurance question and their values were fully imputed.

TABLE 9-1 Item Nonresponse and Imputation Rates for ACS Health Insurance Coverage, 2008 Universe: U.S. Population, All People Who Responded to a 2008 Questionnaire

	All Modes		HU – Mail		HU – CATI/CAPI		GQ	
	Estimate	SE	Estimate	SE	Estimate	SE	Estimate	SE
All People (number in thousands)	300,349	31	153,826	1,073	138,276	1,091	8,247	(x)
Before Editing								
Percentage with complete item response	73.00	0.15	51.80	0.05	96.10	0.03	81.00	0.28
Percentage with at least one but not all response	23.20	0.14	43.80	0.05	1.20	0.02	7.10	0.16
Percentage with no complete items (all nonresponse)	3.80	0.02	4.40	0.02	2.70	0.03	11.80	0.23
After Editing								
Percentage with at least one health insurance type allocated	9.70	0.04	14.70	0.04	3.80	0.03	15.60	0.25

NOTES: SE = standard error.
(x) Rounds to zero.
Subpart h of the question (any other type of health insurance or health coverage plan) is excluded from the nonresponse calculation.
SOURCE: Data from U.S. Census Bureau, 2008 American Community Survey, Turner et al. (2009).

Comparisons to Other National Surveys

Data users and policy makers who have relied on other surveys for estimates of health insurance coverage will be interested in how ACS estimates compare with other sources. The CPS ASEC and the NHIS are both widely used sources of estimates of health insurance coverage, although neither is a gold standard. Both surveys produce estimates that are particular to their own contexts, question wording, and processing regimens.

The following sections briefly describe the surveys, discuss the structure of the health insurance questions, and present coverage estimates from the ACS side-by-side with estimates from the CPS ASEC and the NHIS.

ANNUAL SOCIAL AND ECONOMIC SUPPLEMENT TO THE CURRENT POPULATION SURVEY

Design

The CPS is a monthly survey that the Census Bureau conducts for the Bureau of Labor Statistics to provide data on labor force participation and unemployment. Data on health insurance coverage are collected through the ASEC, which is administered February through April. About 76,200 households are sampled per year. The CPS ASEC sample is the civilian noninstitutionalized population of the United States.[8]

CPS ASEC data are collected through a combination of telephone and personal visit modes using computer-assisted instruments. The Census Bureau publishes CPS ASEC estimates of health insurance coverage for the nation and all states.

Health Insurance Coverage Question

The CPS ASEC income and health insurance coverage questions are asked at the household level, that is, "Does anyone in the household . . . ?" If the answer is "yes," the CPS ASEC goes on to ask "Who . . . ?" This is distinct from the ACS questionnaire, which asks all the questions about each person individually, that is, "Does this person . . . ?" From a cognitive and operational perspective, each approach has benefits and challenges. The CPS ASEC asks respondents to recall their insurance status for the prior calendar year (January through December). Hence, respondents

[8] Members of the armed forces living off post or with their families on post are included if at least one civilian adult lives in the household.

need to recall insurance coverage for a period that began 14 to 16 months prior to the interview. The question series covers a comprehensive list of insurance types that include public program names specific to the state in which the interview is conducted. Finally, if the person does not indicate coverage, a verification question asks specifically about his or her coverage status.[9] The CPS ASEC health insurance question set and editing result in an estimate intended to be of those uninsured for all of the previous calendar year. Previous research has indicated that the long reference period is a limitation of the CPS ASEC methods, with the estimate of the uninsured too high for a "full year" measure and more closely approximating a point-in-time estimate.[10]

NATIONAL HEALTH INTERVIEW SURVEY

Design

The NHIS is an ongoing survey conducted throughout the year by the National Center for Health Statistics to monitor the health of the nation. It has been conducted since 1957. The NHIS consists of a Basic Module, including the Family Core, the Sample Adult Core, and the Sample Child Core, as well as several supplements that vary from year to year. In recent years, slightly less than 35,000 households were interviewed. The NHIS sample is the civilian noninstitutionalized population of the United States.

NHIS data are collected through an in-person survey using computer-assisted interviewing. The sample for the NHIS includes data from the 50 states and the District of Columbia. However, it is not designed to provide state-level estimates; the lowest level of geography publically available is census region.

Health Insurance Coverage Question

Like the ACS, the NHIS asks the respondent about insurance status and coverage type at the time of the survey. The NHIS also asks if the respondent has been uninsured for at least part of the year prior to the interview and if the respondent has been uninsured for more than a year at the time of the interview. The question series includes a comprehensive

[9] For more information on health insurance coverage in the CPS ASEC, see: http://www.census.gov/hhes/www/hlthins/hlthins.html [October 2010].

[10] See appendix C of "Income, Poverty, and Health Insurance Coverage in the United States: 2008" P60-236(RV), for more information about the quality of the CPS ASEC health insurance estimates, available: http://www.census.gov/prod/2009pubs/p60-236.pdf [October 2010].

list of insurance options that include public program names specific to the state in which the interview is conducted, as well as open-ended response options. A verification question is included to confirm that respondents who did not respond that they were enrolled in any insurance program are, in fact, uninsured. The NHIS also edits variables based on supplemental information that the interviewer may collect, such as statements or insurance cards that respondents display.[11]

HEALTH INSURANCE QUESTION DESIGN ISSUES

The ACS is the first major federal mailout-mailback survey to include health insurance questions. The ACS uses a set of health insurance categories that are similar in conceptual scope to other surveys, like the CPS ASEC and the NHIS, but there are methodological differences that highlight the limitations of soliciting health coverage information in a mailout-mailback environment.

The ACS, since it utilizes a paper survey instrument, does not allow the customization of questions to reflect the specific state health programs (or Medicaid/Children's Health Insurance Program funded programs) for which residents of a particular state or locality can apply. The CPS ASEC and the NHIS, in contrast, are conducted entirely through computer-assisted instruments and are able to use state-specific public program names in questions. This mechanism has been shown to help respondents identify their enrollment in public health programs (Eberly et al., 2009). However, at this time the extent to which the lack of state-specific program names biases ACS estimates is uncertain.

Another limitation of a mailed paper survey instrument is the inability to use customized questions or wording for subgroups of the population, such as children. The CPS ASEC and the NHIS, in contrast, utilize specific questions aimed at children to ensure that coverage under the Children's Health Insurance Program is reported.

Although the ACS collects information on a number of important health insurance covariates, such as housing characteristics, public program participation, socioeconomic status, and functional limitations, it lacks a number of covariates that are found in the CPS ASEC and the NHIS. These include general reported health and disability status and detailed income and employment measures and, in the NHIS, health conditions and health care utilization.

The ACS health insurance coverage question uses a clearly defined current coverage measure, referred to as "point in time," that is easily interpreted. In contrast, the CPS ASEC asks respondents to report any

---

[11] For more information on the NHIS, see: http://www.cdc.gov/nchs/nhis.htm.

coverage they had in the preceding calendar year. Numerous studies have found that this question format downwardly biases estimates of coverage. Indeed, the CPS ASEC bias is so severe that it is closer to other surveys' point-in-time measures than it is to other sources of all-year coverage (Congressional Budget Office, 2003; Davern et al., 2007a).

Despite these challenges and deficiencies, the ACS has a number of attributes that benefit data users and researchers. Most notably, its large sample size (about 30 times that of the CPS) and certain selection of all U.S. counties allow it to produce estimates for state and substate geographic areas and for key population subgroups.

COMPARISON OF HEALTH INSURANCE COVERAGE ESTIMATES

In order to compare 2008 ACS data with data from the 2009 CPS ASEC (2008 calendar year estimates) and the 2008 NHIS public-use files, we defined health insurance characteristics in the CPS ASEC and the NHIS as similar to ACS rules.[12] In this way, variables for each of the seven ACS health insurance types were created for the CPS ASEC and the NHIS. These comparisons illustrate how the ACS estimates of the uninsured fit with these other national surveys. The 2008 ACS point-in-time estimates are compared with the CPS ASEC 2008 all-year uninsured estimates and the 2008 NHIS point-in-time estimates. In addition to the uninsured measure, differences in survey design may influence the results. All comparative statements have undergone statistical testing and are significant at the 95 percent confidence level unless otherwise noted.

It is important to note that, unlike the CPS ASEC and the NHIS, the ACS edits for nonresponse did not use a rules-based assignment of health insurance coverage (called consistency or coverage edits). In the ACS, these types of edits are being implemented in the 2009 estimates and are discussed in further detail in the Data Processing section of this paper.

Table 9-2 shows the baseline rates of health insurance coverage from the three surveys. The ACS health insurance coverage rate was 84.9 percent, not statistically different from the NHIS rate of 85.2 percent. This high level of consistency is a good sign for the ACS, which is conceptually similar to the NHIS, as they both measure current coverage. The CPS ASEC health insurance coverage rate was 84.6 percent. Although the statistical test of the difference between the ACS and the CPS ASEC showed evidence of difference, these two estimates do not appear meaningfully different—both round to 85 percent of the population.

Table 9-2 shows that the ACS direct purchase rate is 14.2 percent,

[12] For example, the CPS ASEC estimate of military health care was separated into TRICARE/other military health care and VA coverage.

which is 5.3 percentage points higher than the CPS ASEC and 7.6 points higher than the NHIS. Given that overall levels of health insurance coverage are similar, the direct purchase results suggest that the ACS is classifying direct purchase differently from the alternative surveys, but that at aggregate levels respondents are consistently indentifying that they are covered by some form of coverage. The ACS direct purchase estimate is particularly worrisome, as previous research has shown that the CPS ASEC overestimates administrative totals of the direct purchase population (Cantor et al., 2007). As an all-year measure, the CPS ASEC should theoretically exceed the point-in-time measure of the ACS. As such, the ACS rate is likely to be an overestimate of the direct purchase population. Possible reasons for the higher estimate of direct purchase coverage are discussed in the Direct Purchase section of this paper.

The health insurance coverage rates for children under age 18 are also shown in Table 9-2. Both the ACS and the CPS ASEC estimate that 90.1 percent of children have health insurance coverage, and the NHIS estimates that 91.0 percent of children have health insurance coverage. The difference between the ACS and the NHIS is statistically significant but not meaningful.

The ACS had a higher percentage of children under age 18 with employer-sponsored insurance, 56.2 percent, than the NHIS rate of 54.0 percent. The ACS had a lower percentage of children under age 18 with employer-sponsored insurance than the CPS ASEC rate of 58.9 percent. The ACS estimated a higher proportion of children with direct purchase coverage, 9.2 percent, than the CPS ASEC rate of 5.1 percent or the NHIS rate of 3.4 percent.

The ACS found fewer children under age 18 with coverage from a public means-tested health insurance program, 27.8 percent, than the CPS ASEC rate of 30.3 percent or the NHIS rate of 31.4 percent. This difference may reflect methodological differences in the data collection processes—including the fact that the 2008 ACS, unlike the CPS ASEC and the NHIS, does not include a consistency edit.

Table 9-3 shows uninsured rates by selected demographic and economic characteristics from the three surveys. The uninsured rate for non-Hispanic whites was not statistically different among the surveys: 10.6 percent in the ACS, 10.8 percent in the CPS ASEC, and 10.6 percent in the NHIS.[13] The ACS had a lower percentage of uninsured non-Hispanic

[13] The surveys allow respondents to choose more than one race. Except for the Multiple Race category, race groups discussed in this paper refer to people who indicated only one racial identity among the six major categories: white, black or African American, American Indian and Alaska Native, Native Hawaiian and Other Pacific Islander, and Some Other race. The use of single-race population in this paper does not imply that it is the preferred method of presenting or analyzing data. A variety of approaches are used.

TABLE 9-2 Health Insurance Coverage in the ACS, CPS ASEC, and NHIS, 2008 Universe: U.S. Civilian Noninstitutionalized Population

	ACS		CPS ASEC[a]			NHIS		
	Estimate	SE	Estimate	SE	ACS-CPS Difference	Estimate	SE	ACS-NHIS Difference
All People								
Any Coverage								
Insured	84.90	0.05	84.60	0.11	*	85.20	0.25	
Uninsured	15.10	0.05	15.40	0.11	*	14.80	0.25	
Coverage Type								
Employer	58.70	0.06	58.50	0.15		56.30	0.42	*
Direct purchase	14.20	0.03	8.90	0.08	*	6.60	0.20	*
Medicare	13.50	0.01	14.30	0.10	*	13.90	0.24	
Medicaid	13.40	0.03	14.10	0.10	*	13.90	0.27	
TRICARE/Military	2.50	0.01	2.70	0.05	*	2.10	0.14	*
VA	2.10	0.01	1.20	0.03	*	1.00	0.05	*
IHS[b]	0.50	—	0.30	0.02	*	0.30	0.07	*

People Under Age 18 Years								
Any Coverage								
Insured	90.10	0.06	90.10	0.18		91.00	0.43	*
Uninsured	9.90	0.06	9.90	0.18		9.00	0.43	*
Coverage Type								
Employer	56.20	0.10	58.90	0.29	*	54.00	0.69	*
Direct purchase	9.20	0.05	5.10	0.13	*	3.40	0.30	*
Medicare	0.70	0.01	0.80	0.05	*	0.30	0.05	*
Medicaid	27.80	0.08	30.30	0.27	*	31.40	0.65	*
TRICARE/Military	2.30	0.02	2.60	0.09	*	2.10	0.26	
VA	0.10	0.01	0.40	0.04	*	0.10	0.05	
IHS[b]	0.60	0.01	0.40	0.04	*	0.40	0.14	

NOTES: *Statistically different from zero at the 95 percent confidence level.
— Represents or rounds to zero.
SE = standard error.
[a]The CPS ASEC asks about health insurance coverage over the prior calendar year, however, there is considerable uncertainty as to how respondents answer the health insurance questions in the survey (Congressional Budget Office, 2003; Davern et al., 2007a). It appears that the CPS ASEC, which purports to be a measure of all year uninsured, is closer to a measure of uninsurance at a point in time. The ACS and the NHIS estimates are measures of point-in-time uninsured.
[b]Indian Health Service is not considered coverage for the purpose of tabulating summary coverage measures.
SOURCE: Data from U.S. Census Bureau, 2008 American Community Survey, and the 2009 Annual Social and Economic Supplement to the Current Population Survey; 2008 National Health Interview Survey public-use files; Turner et al. (2009, Table 5).

TABLE 9-3 Uninsured Rate by Selected Characteristics in the ACS, CPS ASEC, and NHIS, 2008 Universe: U.S. Civilian Noninstitutionalized Population

	ACS		CPS ASEC[a]			NHIS		
	Estimate	SE	Estimate	SE	ACS-CPS Difference	Estimate	SE	ACS-NHIS Difference
Percentage of People Uninsured	15.10	0.05	15.40	0.14	*	14.80	0.25	
Age								
Under 6 years	8.60	0.07	8.70	0.30		7.60	0.60	
6 to 18 years	11.20	0.07	11.00	0.24		10.50	0.44	
19 to 64 years	19.90	0.06	20.30	0.18	*	20.00	0.32	
65 years and over	1.40	0.02	1.70	0.10	*	0.60	0.08	*
Sex								
Male	16.60	0.06	17.00	0.18	*	16.40	0.30	
Female	13.60	0.04	13.80	0.15		13.30	0.28	
Race and Hispanic Origin[b]								
White alone, NH	10.60	0.04	10.80	0.14		10.60	0.28	
Black alone, NH	18.00	0.10	19.00	0.41	*	16.40	0.50	*
Asian alone, NH	14.60	0.16	17.60	0.70	*	12.40	0.86	*
AIAN alone, NH	31.50	0.48	25.90	1.50	*	22.80	4.69	
NHOPI alone, NH	16.00	0.83	15.30	2.43		n/a	n/a	
Some other race alone, NH	20.90	0.69	n/a	n/a		n/a	n/a	
Two or more races, NH	13.60	0.22	13.20	0.83		14.60	1.58	
Hispanic or Latino (of any race)	31.50	0.13	30.70	0.44		31.60	0.72	

Income Relative to the Federal Poverty Level								
0-99% FPL	29.10	0.11	30.40	0.47	*	26.50	0.78	*
100-199% FPL	26.00	0.10	24.30	0.37	*	25.70	0.58	
200% FPL	9.70	0.04	10.00	0.13	*	9.50	0.22	
Citizenship Status								
U.S. citizen	12.70	0.04	13.10	0.13	*	12.30	0.24	
Not a U.S. citizen	46.20	0.15	44.70	0.68	*	46.90	1.12	
Marital Status								
Not married	17.60	0.06	18.30	0.18	*	16.80	0.30	*
Married	11.40	0.04	11.20	0.17		12.10	0.30	*

NOTES: *Statistically different from zero at the 95 percent confidence level.
n/a = not available; SE = standard error.
[a]The CPS ASEC asks about health insurance coverage over the prior calendar year, however, there is considerable uncertainty as to how respondents answer the health insurance questions in the survey (Congressional Budget Office, 2003; Davern et al., 2007a). It appears that the CPS ASEC, which purports to be a measure of all year uninsured, is closer to a measure of uninsurance at a point in time. The ACS and the NHIS estimates are measures of point-in-time uninsured.
[b]AIAN = American Indian and Alaska Native, NH = not Hispanic or Latino, NHOPI = Native Hawaiian and Other Pacific Islander.
SOURCE: Data from U.S. Census Bureau, 2008 American Community Survey, and the 2009 Annual Social and Economic Supplement to the Current Population Survey from Lynch et al. (2010); 2008 National Health Interview Survey public-use files.

blacks, 18.0 percent, than the CPS ASEC rate of 19.0 percent, but higher than the NHIS rate of 16.4 percent. Similar to the results for non-Hispanic whites, the uninsured rate for Hispanics was not statistically different among the surveys: 31.5 percent in the ACS, 30.7 percent in the CPS ASEC, and 31.6 percent in the NHIS. In the CPS ASEC and the NHIS, the sample sizes are small for the remaining race categories and are better interpreted when viewed over a longer term. All race and ethnicity categories are presented here for completeness.

The ACS had a slightly lower uninsurance rate for people living in poverty, 29.1 percent, than the CPS ASEC rate of 30.4 percent, but higher than the NHIS rate of 26.5 percent. There are substantial differences in the measurement of income among the three surveys, which may impact the estimates of health insurance coverage by income level. For an in-depth analysis of income data in major federal surveys, see Czajka and Denmead (2008). Although there are some statistically significant differences across other measures examined, those differences tend to be quite small in magnitude.

Table 9-4 shows uninsured rates for children under 19 years by selected demographic and economic characteristics. The uninsured rate by age category is not statistically different among the three surveys. For children under age 6, 8.6 percent were uninsured in the ACS, 8.7 percent in the CPS ASEC, and 7.6 percent in the NHIS. For children aged 6 to 11, 9.7 percent were uninsured in the ACS, 9.2 percent in the CPS ASEC, and 9.0 percent in the NHIS. A similar pattern was seen for children aged 12 to 18. As was true for the overall population, the uninsured estimates from the three surveys are quite consistent for key subgroups of children.

Data Processing

As a whole, the ACS produces estimates of health insurance coverage that are remarkably similar to the CPS ASEC and the NHIS. One potential source of variation could be the different data processing regimens used by the three surveys. Data processing differences in the ACS (consistency edits) and CPS ASEC (imputation bias) are particularly important because there is a high probability that they will change in coming years. Thus, the differences produced in this analysis may not be reflective of differences in future years. In the sections that follow we describe these data processing differences and compare the three surveys after accounting for them.

Consistency Edits

The CPS ASEC and the NHIS employ consistency edits (also called logical coverage edits) that deterministically assign public coverage to people who do not report it. In the CPS ASEC, coverage is assigned to likely enrollees based on a set of rules developed by the Census Bureau in consultation with an independent technical advisory group. The rules apply to Medicare, Medicaid, and military coverage. For example, people over age 65 who report Social Security or Railroad Retirement income are assigned Medicare. This data processing technique is motivated by two assumptions: (1) coverage is generally under reported in surveys, and (2) health insurance is a particularly difficult concept for respondents and is prone to more response error than other socioeconomic concepts. To our knowledge, no study has carefully examined the quality of such edits. However, after consultation with policy and data experts, the Census Bureau determined that these edits reduce the level of error at aggregate population levels.

As mentioned, the 2008 ACS data file has not been edited in this manner. However, in summer 2009, the Census Bureau, in consultation with outside experts, developed a set of edit rules for use in the ACS and will implement them starting with the 2009 ACS (Lynch et al., 2010). The Census Bureau will not retroactively release 2008 data with these edits; however, the State Health Access Data Assistance Center (SHADAC) is in the process of releasing edited microdata through the Minnesota Population Center's Integrated Public Use Microdata Series (IPUMS).[14]

Imputation Bias in the CPS ASEC

The CPS ASEC imputation routine is known to produce less private dependent coverage than is expected from the explicitly reported distribution (Davern et al., 2007b). This problem stems from a rule in the imputation routine that restricts dependent coverage to the immediate family of the policy holder. This is unlike the instrument that allows dependent coverage to be given to any household member. As a result of this rule, imputed dependent coverage is biased downward in reference to the reported distribution. Currently, the Census Bureau is conducting an evaluation of a new imputation routine based on the results of Davern et al. (2007b). The preliminary expectation is that this new routine will be implemented beginning in 2011 for the 2010 estimates.

Data Processing Differences

To account for these data processing differences, we repeat the three-survey comparison using two alternative data sources. To account for the lack of a consistency edit in the 2008 ACS, we used ACS estimates

---

[14] Available at: http://usa.ipums.org/usa [July 2010].

TABLE 9-4 Uninsured Rate for Children by Selected Characteristics in the ACS, CPS ASEC, and NHIS, 2008 Universe: U.S. Civilian Noninstitutionalized Population

	ACS		CPS ASEC[a]			NHIS		
	Estimate	SE	Estimate	SE	ACS-CPS Difference	Estimate	SE	ACS-NHIS Difference
Percentage of Children Under Age 19 Uninsured	10.40	0.07	10.30	0.21		9.50	0.42	*
Age								
Under 6 years	8.60	0.09	8.70	0.30		7.60	0.60	
6 to 11 years	9.70	0.10	9.20	0.30		9.00	0.52	
12 to 18 years	12.50	0.10	12.40	0.30		11.60	0.54	
Race and Hispanic Origin[b]								
White alone, NH	6.90	0.07	7.10	0.22		7.10	0.60	
Black alone, NH	10.40	0.18	11.20	0.61		7.70	0.76	*
Asian alone, NH	9.70	0.30	11.40	1.04		6.80	1.06	*
AIAN alone, NH	27.20	1.09	16.30	2.09	*	13.60	4.85	*
NHOPI alone, NH	11.80	1.94	12.50	5.09		n/a	n/a	n/a
Some other race alone, NH	11.60	1.11	n/a	n/a	n/a	n/a	n/a	n/a
Two or more races, NH	7.40	0.29	6.80	0.82		8.00	1.78	
Hispanic or Latino (of any race)	19.40	0.19	17.90	0.56	*	17.90	0.85	

Income Relative to the Federal Poverty Level								
0-99% FPL	16.70	0.21	16.50	0.63		14.10	1.02	*
100-199% FPL	16.70	0.18	15.30	0.54	*	16.40	0.96	
200% FPL	6.30	0.06	6.40	0.20		5.40	0.34	*
Family Work Status								
No one working in family	13.30	0.28	13.50	0.65		8.90	1.31	*
At least one worker in family	10.20	0.07	9.70	0.21	*	9.60	0.44	

NOTES: *Statistically different from zero at the 95 percent confidence level.

n/a = not available; FPL = federal poverty level; SE = standard error.

[a]The CPS ASEC asks about health insurance coverage over the prior calendar year, however, there is considerable uncertainty as to how respondents answer the health insurance questions in the survey (Congressional Budget Office, 2003; Davern et al., 2007a). It appears that the CPS ASEC, which purports to be a measure of all year uninsured, is closer to a measure of uninsurance at a point in time. The ACS and the NHIS estimates are measures of point-in-time uninsured.

[b]AIAN = American Indian and Alaska Native, NH = not Hispanic or Latino, NHOPI = Native Hawaiian and Other Pacific Islander.

SOURCE: Data from U.S. Census Bureau, 2008 American Community Survey, and the 2009 Annual Social and Economic Supplement to the Current Population Survey public use files; 2008 National Health Interview Survey public-use files.

from Lynch et al. (2010), which are based on 2008 ACS data with the consistency edit. To account for the imputation bias in the CPS ASEC, we used the SHADAC-enhanced CPS ASEC data set.[15] These estimates were produced by removing the fully imputed cases (roughly 10 percent of the entire sample), weighting up the remaining cases to population controls, and adjusting for changes to the health insurance question (State Health Access Data Assistance Center, 2009; Ziegenfuss and Davern, 2010). While this method for correcting the imputation bias in the CPS ASEC may be different from reengineering the imputation routine, we believe it is a close approximation to what would be expected based on the new routine.

In order to understand the contribution of these data processing factors in the differences observed and to provide estimates that may be a better representation of future data, we compare edited ACS, SHADAC-enhanced CPS ASEC, and NHIS data in Table 9-5.

After applying the consistency edits, the ACS uninsured estimate was 14.6 percent compared with 15.1 percent prior to the edits (from Table 9-3). This estimate was nearly identical to the NHIS uninsured estimate of 14.8 percent. The SHADAC-enhanced CPS ASEC uninsured estimate was 14.8 percent, or 0.6 percentage points lower than the CPS ASEC estimate (15.4 percent, from Table 9-3) and identical to the NHIS uninsured estimate.

The ACS (with the consistency edit) had a lower percentage of uninsured non-Hispanic blacks, 17.2 percent, than the SHADAC-enhanced CPS ASEC rate of 18.5 percent but was not statistically different from the NHIS rate of 16.4 percent. Although the uninsured rate was still statistically different for non-Hispanic blacks between the ACS and SHADAC-enhanced CPS ASEC, the estimates were moving closer together. The ACS uninsured rate for non-Hispanic blacks was 18.0 percent prior to the edits and the rate for the CPS ASEC was 19.0 percent (from Table 9-3). This pattern of the estimates moving closer together was seen for most of the other key subgroups.

These results suggest the differences observed are driven in part by the data processing regimens of the alternate surveys. With the application of the consistency edits in the 2009 ACS and the possible implementation of a revised edit routine in the CPS ASEC for 2010 estimates, one would expect to see smaller differences among the surveys in future years.

---

[15] Enhanced data are available from SHADAC's Data Center at: http://www.shadac.org/datacenter [October 2010] and from the Minnesota Population Centers Integrated Public Use Microdata Series (IPUMS) at: http://usa.ipums.org/usa [October 2010].

Comparison of Direct Purchase Estimates

The largest anomaly apparent in Table 9-2 is the relatively high level of direct purchase in the ACS. While these results do not conclusively suggest that the ACS estimate is biased, previous research suggests that the CPS ASEC overcounts the level of direct purchased coverage (Cantor et al., 2007). By extrapolation, the ACS direct purchase estimate is likely to be upwardly biased as well.

The possible misclassification of direct purchase in the ACS could be of two sorts. Respondents could indicate direct purchase in addition to another (and presumably accurate) coverage source. This would mean that misreporting of direct purchase had no effect on the accuracy of other coverage levels. Alternatively, direct purchase could be reported instead of the correct source of coverage. This would result in an undercount of the correct coverage source in addition to the overcount of direct purchase. The source of the direct purchase error in the ACS is currently poorly understood.

Four hypotheses are currently under consideration: (1) the absence of state public program names could result in public program enrollees mistakenly reporting direct purchase; (2) the absence of a qualifier in the response option that explicitly states that direct purchase is not employer-sponsored coverage (as is done in the CPS ASEC) could result in group plan enrollees reporting direct purchase; (3) people with single service plans, such as dental coverage, could be reporting direct purchase; and (4) the person-level roster processing and/or the order of direct purchase in the response list could result in overreporting. Researchers at the Census Bureau in consultation with SHADAC and other interested parties are currently exploring these hypotheses.

A recent study conducted by Urban Institute and SHADAC analysts found larger than expected levels of direct purchase across age, employment, and income distributions, suggesting that the problem is multifaceted and not limited to a single population segment (Lynch and Boudreaux, 2010). The analysis also found that a large portion of the direct purchase population had sociodemographic characteristics that would be consistent with coverage by employer-sponsored insurance or means-tested programs. The researchers logically removed direct purchase coverage from observations that had Medicaid and were eligible for such a benefit, that had military coverage, or that had employer-provided coverage and reported familial employment patterns consistent with employer-provided coverage. Such edits were more conservative than creating a hierarchical variable that assigned direct purchase only to those who reported direct purchase alone. These edits reduced direct purchase coverage among people aged 0 to 64 from 10.5 to 6.6 percent.

TABLE 9-5 Uninsured Rate by Selected Characteristics in the ACS (consistency edits), Enhanced CPS ASEC, and NHIS, 2008 Universe: U.S. Civilian Noninstitutionalized Population

	ACS with Consistency Edits		SHADAC-Enhanced CPS ASEC[a]			NHIS		
	Estimate	SE	Estimate	SE	ACS-CPS Difference	Estimate	SE	ACS-NHIS Difference
Percentage of People Uninsured	14.60	0.05	14.80	0.14		14.80	0.25	
Age								
Under 6 years	8.00	0.07	7.80	0.30		7.60	0.60	
6 to 18 years	10.60	0.07	10.30	0.25		10.50	0.44	
19 to 64 years	19.50	0.06	19.80	0.19		20.00	0.32	
65 years and over	0.90	0.02	1.70	0.11	*	0.60	0.08	*
Sex								
Male	16.10	0.06	16.70	0.19	*	16.40	0.30	
Female	13.20	0.04	13.10	0.15		13.30	0.28	
Race and Hispanic Origin[b]								
White alone, NH	10.30	0.04	10.10	0.14		10.60	0.28	
Black alone, NH	17.20	0.10	18.50	0.44	*	16.40	0.50	
Asian alone, NH	14.20	0.16	16.80	0.73	*	12.40	0.86	*
AIAN alone, NH	29.70	0.47	26.40	1.61	*	22.80	4.69	
NHOPI alone, NH	14.90	0.84	13.60	2.41		n/a	n/a	
Some other race alone, NH	20.30	0.70	n/a	n/a		n/a	n/a	
Two or more races, NH	13.10	0.22	12.60	0.85		14.60	1.58	
Hispanic or Latino (of any race)	30.70	0.13	31.00	0.47		31.60	0.72	

Income Relative to the Federal Poverty Level								
0-99% FPL	27.60	0.11	30.00	0.49	*	26.50	0.78	
100-199% FPL	25.10	0.10	24.10	0.38	*	25.70	0.58	
200% FPL	9.60	0.04	9.30	0.14	*	9.50	0.22	
Citizenship Status								
U.S. citizen	12.30	0.04	12.50	0.13		12.30	0.24	
Not a U.S. citizen	45.50	0.15	46.30	0.72		46.90	1.12	
Marital status								
Not married	17.10	0.06	17.90	0.19	*	16.80	0.30	
Married	10.90	0.04	10.60	0.17		12.10	0.30	*

NOTES: *Statistically different from zero at the 95 percent confidence level.
n/a = not available; FPL = federal poverty level; SE = standard error.
[a]The CPS ASEC asks about health insurance coverage over the prior calendar year, however, there is considerable uncertainty as to how respondents answer the health insurance questions in the survey (Congressional Budget Office, 2003; Davern et al., 2007a). It appears that the CPS ASEC, which purports to be a measure of all year uninsured, is closer to a measure of uninsurance at a point in time. The ACS and the NHIS estimates are measures of point-in-time uninsured.
[b]AIAN = American Indian and Alaska Native, NH = not Hispanic or Latino, NHOPI = Native Hawaiian and Other Pacific Islander.
SOURCE: Data from U.S. Census Bureau and 2008 American Community Survey from Lynch et al. (2010); 2009 SHADAC-Enhanced CPS ASEC; 2008 National Health Interview Survey public-use files.

Given the substantial changes in the individual health insurance market as part of health care reform, improving the ACS instrument or developing postcollection adjustments to the direct purchase estimate will require a close partnership between survey methodologists and policy experts. Under health care reform, it is no longer clear what is meant by direct purchase on a conceptual level. Should people who purchase insurance with premium assistance from a government program or those purchasing coverage through an exchange be counted as direct purchase enrollees? If not, what categories describe them? Once these questions are answered, survey methodologists can begin to craft instruments that are better suited for the postreform environment.

FINAL REMARKS

The ACS, which added a question on health insurance coverage in 2008, is a powerful new resource that can be used both to provide guidance to the implementation of health care reform and to evaluate the impacts of health care reform at the national, state, and local levels. The survey's very large sample size, combined with a sample frame that is representative of all areas of the United States, will support estimates for narrow population subgroups (e.g., young children, adolescents, teenagers transitioning to adulthood) and small geographic areas (e.g., states, counties, communities) that are not possible using other available data sources.

This paper shows that the ACS estimates of health insurance coverage are remarkably consistent with estimates from the other national surveys that are often used to track health insurance coverage—the CPS ASEC and the NHIS. We have found few meaningful differences in estimates of the uninsurance rate in the ACS relative to the CPS ASEC or the NHIS for the overall population or for key population subgroups, including children. This is particularly true after making adjustments for data processing differences across the three surveys. To the extent that such differences are addressed in future rounds of the surveys (e.g., through expected processing enhancements to the ACS and possible improvements in the CPS ASEC imputation routine), we would expect fewer differences in the estimates from the three surveys in the future, ensuring greater consistency in the estimates over time.

Notwithstanding the general consistency of the uninsurance estimates from the ACS with other national surveys, additional work is needed to assess its strengths and weaknesses as more years of data are available over time. For example, more work is needed to understand the relatively high estimate of direct purchase coverage in the ACS relative to other surveys that we have described here. Such work will be critical to establishing the long-term value of the ACS as a resource for policy development, analysis and evaluation, and determining the relative value of the ACS, the CPS ASEC, and the NHIS for alternate uses.

The ACS is particularly promising for state and local analysts and policy makers as they begin to implement health care reform, as it will support estimates for key population subgroups and small geographic areas. The ACS provides data for many states that have not had access to such detailed health insurance coverage information before, as well as for states that may no longer be able to conduct their own household and population surveys in the face of ongoing budget limitations. Ensuring that the ACS fulfills its promise as a state and local resource requires addressing the capacity of state and local analysts to use the ACS data files. It is likely that many states lack the hardware and software capacity to analyze the very large ACS data files. To address this constraint, SHADAC has taken the initiative to provide summary coverage estimates of the ACS on their Data Center online table generator.[16] It is hoped that this user-friendly access point, along with detailed technical documentation and technical assistance, if needed, will help state and local analysts overcome the learning curve related to using the ACS and facilitate rapid policy analysis as states begin addressing the implementation of health care reform.

ACKNOWLEDGMENTS

Much of the work presented here was prepared by the University of Minnesota's State Health Access Data Assistance Center (SHADAC) under contract to the U.S. Census Bureau and presented in the working paper "A Preliminary Evaluation of Health Insurance Coverage in the 2008 American Community Survey," released September 22, 2009. That working paper was led by Joanna Turner while she was at the U.S. Census Bureau and coauthored by Michel Boudreaux (SHADAC) and Victoria Lynch (The Urban Institute) and is available at: http://www.census.gov/hhes/www/hlthins/data/acs/2008/2008ACS_healthins.pdf. Turner joined SHADAC in March 2010. The views expressed are those of the authors and not necessarily those of the U.S. Census Bureau.

[16] Available: http://www.shadac.org/datacenter [October 2010] and http://www.shadac.org/content/acs-info-and-resources [October 2010].

REFERENCES

Cantor, J., Monheit, A.C., Brownlee, S., and Schneider, C. (2007). The adequacy of household survey data for evaluating the nongroup health insurance market. *Health Services Research, 42*(4), 1739-1757.

Congressional Budget Office. (2003). *How Many People Lack Health Insurance and for How Long?* Washington DC: U.S. Congress. Available: http://www.cbo.gov/doc.cfm?index=4210 [October 2010].

Czajka, J., and Denmead, G. (2008). *Income Data for Policy Analysis: A Comparative Assessment of Eight Surveys Final Report*. Report from Mathematica Policy Research, Inc. Available: http://www.mathematica-mpr.com/publications/PDFs/incomedata.pdf [October 2010].

Davern, M., Davidson, G., Ziegenfuss, J., Jarosek, S., Lee, B., Yu, T., Beebe, T.J., Call, K.T., and Blewett, L.A. (2007a). A comparison of the health insurance coverage estimates from four national surveys and six state surveys: A discussion of measurement issues and policy implications. Minneapolis: University of Minnesota. Available: http://www.shadac.org/files/shadac/publications/ASPE_FinalRpt_Dec2007_Task7_2_rev.pdf [October 2010].

Davern, M., Rodin, H., Blewett, L.A., and Call, K.T. (2007b). Are the current population survey uninsurance estimates too high? An examination of the imputation process. *Health Services Research, 42*(5), 2038-2055.

Eberly, T., Pohl, M.B., and Davis, S. (2009). Undercounting Medicaid enrollment in Maryland: Testing the accuracy of the Current Population Survey. *Population Research and Policy Review, 28*(2), 221-236.

Lynch, V., and Boudreaux, M. (2010). Health Insurance Estimates from the ACS: An Analysis of Directly Purchased Coverage. Presentation at the 2010 Joint Meetings of the American Statistical Association, August 3, Miami Beach, FL.

Lynch, V., Boudreaux, M., and Davern, M. (2010). *Applying and Evaluating Logical Coverage Edits to Health Insurance Coverage in the American Community Survey*. U.S. Census Bureau, housing and household economic statistics division. Available: http://www.census.gov/hhes/www/hlthins/publications/coverage_edits_final.pdf [October 2010].

State Health Access Data Assistance Center. (2005). *Reclassifying Health Insurance Coverage for the Indian Health Service in the Current Population Survey: Impact on State Uninsurance Estimates*. Issue Brief #11. Minneapolis: University of Minnesota. Available: http://www.shadac.org/files/IssueBrief11.pdf [October 2010].

State Health Access Data Assistance Center. (2009). *Historical Changes in Current Population Survey Health Insurance Coverage Items for Survey Years 1988 Through 2009*. Issue Brief #19. Minneapolis: University of Minnesota. Available: http://www.shadac.org/files/shadac/publications/IssueBrief19.pdf [October 2010].

Turner, J., Boudreaux, M., and Lynch, V. (2009). *A Preliminary Evaluation of Health Insurance Coverage in yhe 2008 American Community Survey*. U.S. Census Bureau, housing and household economic statistics division. Available: http://www.census.gov/hhes/www/hlthins/data/acs/2008/2008ACS_healthins.pdf [October 2010].

Ziegenfuss, J., and Davern, M. (2010). Twenty years of coverage: An enhanced Current Population Survey—1989-2008. *Health Services Research*, doi: 10.1111/j.1475-6773.2010.01171.x. Available: http://onlinelibrary.wiley.com/doi/10.1111/j.1475-6773.2010.01171.x/full [October 2010].

10

Income and Poverty Measurement in Surveys of Health Insurance Coverage

John L. Czajka
Mathematica Policy Research, Inc.

Each year the Census Bureau produces a set of state-level estimates of children under age 19 in families with incomes at or below 200 percent of the federal poverty level and the subset of these who are without health insurance coverage. The estimates, which are derived from the Annual Social and Economic Supplement (ASEC) to the monthly Current Population Survey (CPS), were mandated by Congress for the purpose of allocating federal funds for matching state expenditures under the Children's Health Insurance Program (CHIP), but they have become a critical source of data for monitoring the states' progress in reducing the number of low-income uninsured children. Because the CPS ASEC samples for most states are not large enough to support estimates of low-income uninsured children at satisfactory levels of precision, the Census Bureau combines 3 years of data to produce the estimates for each year. Even with 3 years of data, however, the estimates for many states are still not as precise as might be desired, and the need to combine 3 years of data makes the estimates less sensitive to year-to-year change in the number of low-income uninsured children. While this reduced sensitivity may have merits for funding allocation, it limits the usefulness of the estimates for monitoring state levels of health insurance coverage among low-income children.

With the recent reauthorization of CHIP, Congress made provision for potential enhancements to the annual estimates of low-income uninsured children while simultaneously ending their role in CHIP funding allocation. One of the options that the law mandates the Census Bureau to consider is to supplement or replace the CPS estimates with alternative estimates drawn from the American Community Survey (ACS)—a relatively new Census Bureau survey with an annual sample that is 30 times that of the CPS ASEC supplement. With its ability to support not only state, but also substate estimates, the ACS would appear particularly well suited to serve as the principal source of data for monitoring the number of low-income uninsured children. For this reason, measures of health insurance coverage were added to the ACS questionnaire in 2008.

Although much attention is being focused on the quality of these new estimates of health insurance coverage as a major factor in the choice between the CPS or the ACS, the quality of the income data collected in the ACS may be as important in determining its ultimate viability as a source of annual estimates of low-income children and low-income uninsured children. Arguably, the measurement of income is more challenging than the measurement of health insurance coverage, which requires only binary responses as opposed to dollar amounts. For income measurement, the CPS holds clear advantages over the ACS. For one, the CPS is the official source of annual estimates of income and poverty in the United States, whereas the ACS collects comparatively little information on personal income and does so through a questionnaire that two-thirds of the respondents complete without the assistance of an interviewer and submit by mail. Given these potential limitations, do the ACS income data measure up to the CPS sufficiently well to warrant their use in producing the mandated annual estimates of low-income children and low-income uninsured children?

Policy analysts also have reason to be interested in the ACS as a potential data source for a wide variety of analyses of state and local variation in health insurance coverage, and income is likely to play a key role in many such analyses. For such applications, the demands placed on the ACS income data are likely to exceed those that must be met in providing satisfactory estimates of low-income children.

This paper compares estimates of income obtained from the ACS and the CPS. Selected comparisons include a third survey as well—the National Health Interview Survey (NHIS), which is sponsored and designed by the National Center for Health Statistics (NCHS) with fieldwork conducted by the Census Bureau. In certain respects the design of the NHIS resembles that of the ACS, with interviews conducted on a rolling basis throughout the calendar year. The measurement of health insurance coverage is a major element of the NHIS, but the measurement of income is much more limited than even that in the ACS. Two other surveys, the Survey of Income and Program Participation (SIPP) and the Household Component of the Medical Expenditure Panel Survey (MEPS) also collect data on both income and health insurance coverage, but their longitudinal designs are less well suited to monitoring the number of low-income uninsured children than those of the CPS, the ACS, or the NHIS, and, like the NHIS, their samples are not designed for state-level estimation. For this reason, this paper does not include comparative estimates from these other surveys.

Following an overview of income measurement and other features that differentiate among the three surveys, I present findings from two empirical analyses comparing the surveys. The first set of findings is drawn from a recent, comprehensive, and systematic assessment of the income data in eight major surveys and their utility for policy-related analyses. This research was conducted by Mathematica Policy Research, Inc., and its subcontractor, Denmead Services & Consulting, under a contract with the Office of the Assistant Secretary for Planning and Evaluation in the U.S. Department of Health and Human Services.[1] The second set of findings, which is restricted to the CPS and the ACS, was prepared to examine specific issues related to the use of income data in the analysis of health insurance coverage.[2] The paper closes with a brief assessment of the implications of these findings for applications of income and health insurance data from the three surveys.

INCOME MEASUREMENT IN HOUSEHOLD SURVEYS

Despite its wide use in analysis of social and economic phenomena and its critical role in policy analysis, income is difficult to measure well in household surveys. Surveys with a major focus on income devote hundreds of questions to its measurement, whereas surveys that emphasize other topics may devote just a few questions to capturing income. Ideally, those who design the income modules of surveys would be able to draw on established practice to determine how to obtain with a given number of questions the best all-around measure of income or the best measure focusing on a specific application. Such is not the case, however. The most rigorous study of income measurement conducted in the United States to date, by the Income Survey Development Program, was focused on the collection of comprehensive data on income and laid the cornerstone for the design of the SIPP, but it did not shed much light on how a survey could most effectively get by with less. Indeed, the Mathematica study suggests that the SIPP approach, with its focus on monthly income, may be less effective than simpler approaches to measuring, say, annual earnings or even total annual income. Below I provide a brief overview of income measurement in the CPS, the ACS, and the NHIS and then discuss

[1] Czajka and Denmead (2008).

[2] Some of these findings were presented at the annual meeting of the American Association for Public Opinion Research, May 13-16, 2010, Chicago.

other differences among the surveys that have implications for the respective survey estimates of income and poverty.

DIFFERENCES IN THE MEASUREMENT OF INCOME ACROSS SURVEYS

As the official source of annual income and poverty estimates for the United States, the CPS in a sense defines how these constructs should be measured. Thus one can refer to the CPS income concept (pretax money income as defined in the survey), the CPS family (two or more persons living in the same household and related by blood, marriage, or adoption), and CPS residence rules (which include in a household any usual members who are temporarily living elsewhere). The CPS collects data on the presence of more than 50 sources of income and captures up to 24 annual dollar amounts for each sample member age 15 and older. The reported incomes of individual family members at the time of the interview are summed to obtain a measure of total family income for the preceding calendar year.

By contrast, the ACS collects income for up to eight sources for each sample person age 15 and older, combining many sources for which the CPS collects separate reports. The eight sources listed in the questionnaire are

1. wages, salary, commissions, bonuses, or tips from all jobs (prior to deductions);
2. self-employment income from own nonfarm businesses or farm businesses, including proprietorships or partnerships;
3. Interest, dividends, net rental income, royalty income, or income from estates and trusts;
4. Social Security or Railroad Retirement;
5. Supplemental Security Income;
6. any public assistance or welfare payments from the state or local welfare office;
7. retirement, survivor, or disability pensions; and
8. any other sources of income received regularly such as veterans' payments, unemployment compensation, child support, or alimony (not to include lump-sum payments).

This last source serves as a catchall for individual sources that are not explicitly mentioned in the ACS but are identified explicitly in the CPS questionnaire. In addition, each of the preceding sources is captured by multiple questions in the CPS.

CPS interviews are conducted in person or by telephone using computer-assisted personal interviewing (CAPI) or computer-assisted telephone interviewing (CATI). By contrast, about two-thirds of the ACS responses are collected by mail, using a mailout-mailback questionnaire. Most of the remaining responses are collected by CAPI, with the remainder being collected by CATI. The CATI/CAPI responses are collected from a sample of nonrespondents to the mail questionnaire. Unlike the CPS, responding to the ACS is mandatory by law, and the Census Bureau obtains an overall response rate of about 97 percent (weighted to reflect the sampling of nonrespondents to the mail questionnaire). The response rate for the monthly labor force component of the CPS reaches the low 90s, but an additional 8 percent or more of the respondents to the labor force questionnaire do not complete the ASEC supplement interview.

The NHIS collects total family income from a single question asked of the family respondent. Earnings from employment are collected from all persons aged 18 and older, but this is separate from and not reconciled with reported family income. The NHIS interviews are conducted in person using CAPI, and they make extensive use of flash cards—but not for income.

Both the CPS and the NHIS ask their respondents to report their income for the previous calendar year. In the ACS, respondents are asked to report their income for the past 12 months. For persons completing the ACS at the beginning of the year, the past 12 months are January through December of the prior year; for persons completing the survey at the end of the year, the past 12 months are December of the prior year through November of the current year. Thus, the income data collected from ACS households during a given calendar year span a 23-month period centered on December of the prior year.[3] To convert the ACS income to a common reference period—specifically, the calendar year in which the data were collected—the Census Bureau applies an inflation adjustment, defined as the ratio of the average monthly price index for the survey year to the average index for the reference period. These monthly adjustment factors are used internally and are applied to published estimates, but the ACS public-use file contains only an average adjustment factor for the 12 survey months, because the Census Bureau has elected not to reveal the survey month. To calculate income relative to poverty, the Census Bureau adjusts the poverty thresholds rather than the reported income. That is, the income reported for a given reference period is divided by the average monthly

[3] Income reported in ACS published data and online tables that are based on internal files is adjusted across the rolling reference period to the same real dollars, based on the consumer price index. Income in ACS public-use files is not adjusted for inflation, although an average inflation factor is provided.

threshold for that reference period. The estimates of income relative to poverty derived in this manner are reported on the public-use file.

OTHER DIFFERENCES AMONG THE SURVEYS

In addition to these differences in the measurement of income, there are several additional differences that users need to take into account.

First, the surveys represent populations at different times. The CPS ASEC is conducted primarily in March, with additional interviews in February and April. The survey is weighted to represent the population as of March 1.[4] Both the ACS and the NHIS represent an average of populations over a calendar year. The ACS is weighted to July 1 population totals, and the four segments of the NHIS are weighted, separately, to February 1, May 1, August 1, and November 1 of the survey year, then combined to create a single annual weight on the public-use file, with an effective reference date of mid-June.

Second, the CPS includes in the household any college students living away from home, whereas the NHIS and the ACS do not. The NHIS samples and interviews students independently and conducts interviews in dormitories to collect data from students in campus housing, which the CPS does not do. The ACS follows decennial census practice in counting college students where they live at the time of the interview, although it did not begin to collect data from residents of dormitories and other group quarters until 2005.[5]

Third, both the CPS and the ACS define the family unit to include only those persons residing in the same housing unit who are related by blood, marriage, or adoption. The NHIS family definition is more expansive, including unmarried partners (and any relatives living with them) and foster children in the same family. Differences in the residence rules and the family definition affect the number and composition of CPS, ACS, and NHIS families. This, in turn, has implications for the measurement of poverty, which is defined at the family level.

Fourth, while employing the same family definition as the CPS, the ACS does not collect data on relationships among persons who are unrelated to the householder. Thus, if a household includes the householder

[4] CPS ASEC interviews are conducted in February, March, and April in the week that includes the 19th. Historically, the annual income supplement was conducted solely in March, so a March-based weight was appropriate. The population estimates that are used as controls in the CPS and other surveys have a reference data corresponding to the first of each month as key components of these population estimates—births and deaths—are collected by calendar month.

[5] Until 2005, students living in dormitories were excluded from the ACS universe, and the population totals used in weighting the survey were adjusted accordingly.

and a husband, wife, and child who are all unrelated to the householder, the ACS would report this as a householder and three unrelated individuals, and none of the relationships among the three family members would be recorded.

SURVEY ESTIMATES OF INCOME AND POVERTY, 2002

The estimates reported in this section, for calendar year 2002, were adjusted in order to make the estimates as comparable as possible. Under the heading Adjustments to Enhance Comparability, the adjustments that were applied to the surveys included in this paper are described. Total Income compares survey estimates of total income, both in the aggregate and by quintile of family income. The next section examines earned and unearned income aggregates and quintiles. Under Poverty, survey estimates of the number of persons and the fraction of the population classified as poor and near-poor among the population as a whole and among children, nonelderly adults, and the elderly are compared. It also provides estimates of the impact of the NHIS family definition on measured poverty. Finally, Aspects of Data Quality examines alternative measures of income data quality, including allocation of missing data due to item nonresponse and the prevalence of rounding.

Adjustments to Enhance Comparability

The adjustments that were applied to the eight surveys covered by Czajka and Denmead (2008) were designed to make the estimates comparable in terms of income reference period, universe, income concept, and family concept. Most of the adjustments addressed features of the surveys other than the three covered here. I mention adjustments that affected the CPS, the ACS, and the NHIS. ACS income data were multiplied by the average adjustment factor described above. Because of differential inclusion and weighting of members of the active-duty armed forces, all families including such members were removed. Unrelated children under age 15 were also removed, as they are excluded from official estimates of poverty. The most complex adjustment involved the creation of CPS families within the subset of NHIS families that included members based on the broader NHIS family concept—specifically, unmarried partners and their children as well as foster children. In each case, the family members were reassembled into two or more CPS families, and the income of the original family was apportioned among the new families. I did not attempt to adjust for differences in survey timing, which means that the NHIS estimates will reflect a slightly larger population than the CPS estimates, and both surveys' estimates will reflect a larger population than the ACS estimates.

Total Income

As a summary statistic, total or aggregate income is appealing for its simplicity and its use of all of the income data collected in each survey. However, aggregate income is heavily dependent on the amount of income captured from the upper end of the income distribution, which holds the least interest for policy analysis. In presenting estimates of aggregate income, I include a breakdown by quintile of family income, which makes it possible to compare the surveys with respect to their collection of income from different segments of the income distribution. Table 10-1 presents estimates of aggregate income for the whole population and by quintile of family income for the three surveys.[6] It also shows these amounts as a percentage of the corresponding amounts for the CPS. There is no gold standard for estimates of income, nor do I mean to suggest that the CPS estimates are the best. But because the CPS is the official source of household income and poverty statistics for the United States, expressing other survey estimates of income as a percentage of the CPS provides a useful standardization.

Aggregate income ranges from $6.12 trillion in the NHIS to $6.47 trillion in the CPS—a spread of just 5 percent. That is, with a single question the NHIS captures 95 percent as much total income as the CPS; with a simple instrument primarily filled out by respondents rather than a trained interviewer, the ACS captures 98 percent as much total income as the CPS.

In examining income by quintile, one finds that the ACS aggregates lie within a percentage point of the CPS aggregates (both above and below) through the first three quintiles before dropping to 98 and 97 percent, respectively, of the CPS in the fourth and fifth quintiles. The NHIS compares least favorably in the bottom quintile, capturing only 85 percent as much total income as the CPS (and the ACS). This fraction rises to a peak of 98 percent in the fourth quintile before falling back to 94 percent in the top quintile.

Aggregates in the top quintile may be affected by outliers and by differences in survey practice with respect to the topcoding of public-use data. For example, the CPS assigns the means of topcoded values as their respective topcodes, which preserves overall means and totals, but not all surveys do this for all income items. For this reason, the survey aggregates are summed through the bottom four quintiles. Through four quintiles, the ACS captures 99.1 percent as much total income as the CPS, and the NHIS improves only marginally to 95.3 percent.

[6] In each survey, each of the five quintiles contains the same number of people (weighted) except when the numbers are affected by heaping at quintile boundaries.

TABLE 10-1 Aggregate Income by Quintile of Family Income: Three Surveys

Income Estimate	CPS	ACS	NHIS
	Billions of Dollars		
Aggregate Income, All Persons	6,468.40	6,346.30	6,116.20
Family Income Quintile			
Lowest	370.50	368.70	313.70
Second	774.10	778.40	717.70
Third	1,090.20	1,087.40	1,058.40
Fourth	1,446.80	1,415.80	1,420.70
Highest	2,786.70	2,696.00	2,605.80
Sum Through Four Quintiles	3,681.70	3,650.30	3,510.40
	Percentage of CPS		
Aggregate Income, All Persons	100.00	98.10	94.80
Family Income Quintile			
Lowest	100.00	99.50	84.70
Second	100.00	100.60	92.70
Third	100.00	99.70	97.10
Fourth	100.00	97.90	98.20
Highest	100.00	96.70	93.50
Sum Through Four Quintiles	100.00	99.10	95.30

SOURCE: Mathematica Policy Research, Inc., from tabulations of calendar year 2002 income from the 2003 CPS ASEC supplement; the 2003 NHIS; and prior 12 months income, inflation-adjusted to calendar year 2002, from the 2002 ACS.

Because the CPS and the NHIS are weighted to population totals almost a year later than the ACS, the ACS is at a comparative disadvantage for the measurement of aggregate income. One can compensate by putting aggregate income on a per capita basis. Table 10-2 compares the three surveys with respect to per capita income by quintile. By this measure, the ACS captures 99.8 percent as much total income as the CPS, and the NHIS, which is weighted to slightly larger population totals than the CPS, declines to 94.2 percent of the CPS total. ACS per capita income equals or exceeds CPS per capita income through the first three quintiles and is within 0.4 percentage points in the fourth quintile and within 1.6 percentage points in the top quintile. Putting income on a per capita basis has an uneven impact on the NHIS estimates, which implies that the weighted NHIS sample is not distributed as uniformly by quintile as the other two surveys. A likely explanation is discussed below.

TABLE 10-2 Average Income Per Capita by Quintile of Family Income: Three Surveys

Income Estimate	CPS	ACS	NHIS
	Billions of Dollars		
Aggregate Income, All Persons	22,893.00	22,854.00	21,558.00
Family Income Quintile			
Lowest	6,513.00	6,526.00	5,528.00
Second	13,789.00	14,259.00	12,649.00
Third	19,293.00	19,576.00	18,493.00
Fourth	25,604.00	25,496.00	25,151.00
Highest	49,316.00	48,543.00	46,114.00
	Percentage of CPS		
Aggregate Income, All Persons	100.00	99.80	94.20
Family Income Quintile			
Lowest	100.00	100.20	84.90
Second	100.00	103.40	91.70
Third	100.00	101.50	95.90
Fourth	100.00	99.60	98.20
Highest	100.00	98.40	93.50

SOURCES: Mathematica Policy Research, Inc., from tabulations of calendar year 2002 income from the 2003 CPS ASEC supplement; the 2003 NHIS; and prior 12 months income, inflation-adjusted to calendar year 2002, from the 2002 ACS.

Earned and Unearned Income

Economists divide income into earned and unearned. Earnings include wages and salaries plus self-employment income. Earnings account for 82.8 percent of total income in the CPS, 82.1 percent in the ACS, and 86.0 percent in the NHIS (see Table 10-3).[7]

The similarity between the ACS and the CPS hides a difference in the composition of earnings in the two surveys. The ACS captures 4 percent less wage and salary income than the CPS but 19 percent more self-employment income, which raises the ACS earned income to 97.3 percent of the CPS total. The ACS also captures slightly more (2.2 percent) unearned income than the CPS, which contributes to an overall total income that is 98.1 percent of the CPS total. The NHIS does not collect unearned income, but the difference between total income and earned income collected in the NHIS implies unearned income that is 77 percent

[7] Recall that total family income and earnings by individual family members are measured independently in the NHIS and not reconciled.

TABLE 10-3 Contribution of Earned and Unearned Income to Total Income: Three Surveys

Income Estimate	CPS	ACS	NHIS
	Billions of Dollars		
Total Income	6,468.40	6,346.30	6,118.20
Earned Income	5,354.30	5,207.90	5,261.40
Wages and salaries	5,026.30	4,817.20	n/a
Self-employment	328.00	390.70	n/a
Unearned Income	1,114.10	1,138.30	854.80
	Percentage of Total Income		
Total Income	100.00	100.00	100.00
Earned Income	82.80	82.10	86.0
Wages and salaries	77.70	75.90	n/a
Self-employment	5.10	6.20	n/a
Unearned Income	17.20	17.90	14.0
	Percentage of CPS Income by Source		
Total Income	100.00	98.10	94.60
Earned Income	100.00	97.30	98.30
Wages and salaries	100.00	95.80	n/a
Self-employment	100.00	119.10	n/a
Unearned Income	100.00	102.20	76.70

NOTE: n/a = not available.
SOURCES: Mathematica Policy Research, Inc., from tabulations of calendar year 2002 income from the 2003 CPS ASEC supplement; the 2003 NHIS; and prior 12 months income, inflation-adjusted to calendar year 2002, from the 2002 ACS.

of the CPS total. This implied shortfall is simply an indication that the NHIS does not do as well in obtaining total income with its single question as it does in collecting earned income from all adults.[8]

Comparing survey estimates of earned income by quintile of family income, one finds, interestingly, that both the ACS and the NHIS have more earnings in the lowest quintile of family income than does the CPS (see Table 10-4). The additional earnings range from 12 to 17 percent of the CPS total. The ACS has progressively less total earnings relative to the CPS as the quintile increases. The NHIS, in contrast, has progressively more aggregate earnings relative to the CPS over quintiles two through four.

[8] In the 2002 NHIS, more than a fifth of individuals were in families with more reported total earnings than total family income. Each of these families would have implied negative unearned income.

TABLE 10-4 Aggregate Earned Income by Quintile of Family Income: Three Surveys

Income Estimate	CPS	ACS	NHIS
	Billions of Dollars		
Aggregate Earned Income	5,354.30	5,207.90	5,261.40
Family Income Quintile			
Lowest	176.10	206.50	196.40
Second	542.90	565.30	514.40
Third	889.20	878.80	888.20
Fourth	1,255.90	1,225.50	1,301.90
Highest	2,490.20	2,332.00	2,360.50
Sum Through Four Quintiles	2,864.10	2,876.00	2,900.90
	Percentage of CPS		
Aggregate Income, All Persons	100.00	97.30	98.30
Family Income Quintile			
Lowest	100.00	117.30	111.60
Second	100.00	104.10	94.70
Third	100.00	98.80	99.90
Fourth	100.00	97.60	103.70
Highest	100.00	93.60	94.80
Sum Through Four Quintiles	100.00	100.40	101.30

SOURCES: Mathematica Policy Research, Inc., from tabulations of calendar year 2002 income from the 2003 CPS ASEC supplement; the 2003 NHIS; and prior 12 months income, inflation-adjusted to calendar year 2002, from the 2002 ACS.

Unearned income does not show such clear patterns. Overall, the ACS has slightly more unearned income than the CPS, but, unlike earned income, the ACS has progressively more than the CPS as the quintile rises (see Table 10-5). In the top quintile, the ACS has 23 percent more unearned income than the CPS. With its unearned income measured as a residual rather than a reported amount, the NHIS is erratic. The difference between aggregate total and aggregate earned income is as low as 60 percent of the CPS aggregate in one quintile and as high as 88 percent (in the adjacent quintile).

Moving from dollars to receipt, which the NHIS collects for a number of income sources, I observe striking differences among the surveys with respect to programs that serve low-income families. Both the ACS and the NHIS find a much greater incidence of food stamp and welfare receipt among higher income families than does the CPS (see Table 10-6). While the ACS identifies 9 to 10 percent more participants than the CPS in the bottom two quintiles, this proportion rises sharply through the top three quintiles. In the top quintile the ACS finds nearly four times as

TABLE 10-5 Aggregate Unearned Income by Quintile of Family Income: Three Surveys

Income Estimate	CPS	ACS	NHIS
	Billions of Dollars		
Aggregate Unearned Income	1,114.10	1,138.30	854.80
Family Income Quintile			
Lowest	194.40	162.20	117.30
Second	231.20	213.10	203.30
Third	201.00	208.60	170.20
Fourth	190.90	190.30	118.70
Highest	296.50	364.00	245.30
Sum Through Four Quintiles	817.60	774.30	609.50
	Percentage of CPS		
Aggregate Income, All Persons	100.00	102.20	76.70
Family Income Quintile			
Lowest	100.00	83.40	60.30
Second	100.00	92.20	88.00
Third	100.00	103.80	84.60
Fourth	100.00	99.70	62.20
Highest	100.00	122.80	82.70
Sum Through Four Quintiles	100.00	94.70	74.50

NOTE: Unearned income is the difference between total income, reported in Table 10-1, and earned income, reported in Table 10-4.
SOURCES: Mathematica Policy Research, Inc., from tabulations of calendar year 2002 income from the 2003 CPS ASEC supplement; the 2003 NHIS; and prior 12 months income, inflation-adjusted to calendar year 2002, from the 2002 ACS.

many participants as the CPS (a difference of 700,000 persons). The NHIS is not as extreme as the ACS, but it still has nearly twice the incidence of food stamp and welfare receipt in the top quintile as does the CPS. In the bottom quintile the NHIS compares closely to the ACS. Given the target populations for these two programs, one suspects that the substantially higher receipt of food stamps and welfare in the top quintiles in the ACS and the NHIS relative to the CPS is more indicative of problems with the former two surveys than the CPS. While all three surveys show the expected steep decline in receipt with rising family income, it is likely that both the ACS and the NHIS have far too many high-income families with food stamps or welfare. Without further study, however, it is not clear that the problem necessarily lies with the survey data on receipt as opposed to family income.

TABLE 10-6 Persons in Families with Welfare and/or Food Stamps by Quintile of Family Income: Three Surveys

Income Estimate	CPS	ACS	NHIS
	Thousands of Persons		
All Participants	20,496.00	24,325.00	21,990.00
Family Income Quintile			
Lowest	13,562.00	14,879.00	14,783.00
Second	4,461.00	4,854.00	4,355.00
Third	1,748.00	2,396.00	1,685.00
Fourth	493.00	1,273.00	719.00
Highest	233.00	923.00	447.00
Sum Through Four Quintiles	20,263.00	23,402.00	21,543.00
	Percentage of CPS		
Aggregate Income, All Persons	100.00	118.70	107.30
Family Income Quintile			
Lowest	100.00	109.70	109.00
Second	100.00	108.80	97.60
Third	100.00	137.00	96.40
Fourth	100.00	258.40	145.90
Highest	100.00	396.80	192.20
Sum Through Four Quintiles	100.00	115.50	106.30

SOURCE: Mathematica Policy Research, Inc., from tabulations of the 2003 CPS ASEC and the 2003 NHIS.

Poverty

Another useful summary statistic, but one that is informative about only the lower end of the income distribution, is the poverty rate—that is, the percentage of persons whose family incomes lie below the official poverty threshold. Estimates of the number of poor and near-poor (defined as those between 100 and 200 percent of the poverty threshold) are important measures for policy analysis.[9] Marked differences across surveys in estimates of the poor and near-poor would be a source of concern among policy analysts and other data users. They would imply, for example, that the poor in one survey do not represent the same people as the poor in another survey.

Estimates of the Poor and Near-Poor

The CPS and the ACS compare closely, with poverty rates of 12.2 percent and 12.5 percent, respectively (see Table 10-7). The NHIS, in contrast,

[9] Near-poor does not have a standard definition. I use the term to give a name to those with low income (below 200 percent of poverty) but not poor. Elsewhere, near-poor is sometimes used to identify persons between 100 and 125 percent of poverty.

TABLE 10-7 Estimates of the Poor and Near-Poor: Three Surveys

Estimate	CPS	ACS	NHIS
	Millions of Persons		
All Persons	282.55	277.69	283.71
Poverty Status			
Poor	34.38	34.61	41.58
Near-poor	51.81	49.28	53.91
Total Low Income	86.19	83.89	95.49
	Percentage of the Population		
All Persons	100.00	100.00	100.00
Poverty Status			
Poor	12.20	12.50	14.70
Near-poor	18.30	17.70	19.00
Total Low Income	30.50	30.20	33.70

NOTES: The poor have a family income below the poverty threshold. The near-poor have a family income at or above the poverty threshold but below twice the poverty threshold.
SOURCES: Mathematica Policy Research, Inc., from tabulations of poverty status in calendar year 2002 from the 2003 CPS ASEC supplement and the 2003 NHIS, and poverty status in the prior 12 months, inflation-adjusted to calendar year 2002, from the 2002 ACS.

is an outlier with an estimate of 41.6 million poor and a poverty rate of 14.7 percent. The NHIS poverty rate with a CPS family definition is more than 2 percentage points higher than those of the other two surveys.

Combining the estimates of the poor and near-poor, which define the low-income population, the ACS assigns a slightly smaller fraction of the population to that status than the CPS: 30.2 versus 30.5 percent. The NHIS has a larger fraction—33.7 percent of the population or 95.5 million persons—to have low income. The number of persons estimated to have low income in the NHIS exceeds the CPS estimate by 9.3 million.

Estimates of Poor and Near-Poor Children, Nonelderly Adults, and Elderly

The CPS and ACS estimates of children in low-income families are very similar at about 27.4 million, or 38.2 to 38.8 percent of the child population (see Table 10-8). The NHIS has somewhat more at 29.7 million or 41.4 percent. The NHIS also has the most poor children, with a child poverty rate that exceeds the other surveys by 2 to 3 percentage points, but it has no more near-poor children than the CPS. In fact, the estimates of near-poor children across the three surveys vary from only 14.9 to 15.4 million or 21.1 to 21.5 percent of the child population.

TABLE 10-8 Estimates of Poor and Near-Poor Children: Three Surveys

Estimate	CPS	ACS	NHIS
	Millions of Persons		
All Children Under Age 18	71.67	70.79	71.73
Poverty Status			
Poor	12.03	12.51	14.29
Near-poor	15.38	14.94	15.41
Total Low Income	27.41	27.45	29.70
	Percentage of the Population		
All Children Under Age 18	100.00	100.00	100.00
Poverty Status			
Poor	16.80	17.70	19.90
Near-poor	21.50	21.10	21.50
Total Low Income	38.20	38.80	41.40

NOTES: The poor have a family income below the poverty threshold. The near-poor have a family income at or above the poverty threshold but below twice the poverty threshold.
SOURCES: Mathematica Policy Research, Inc., from tabulations of poverty status in calendar year 2002 from the 2003 CPS ASEC supplement and the 2003 NHIS, and poverty status in the prior 12 months, inflation-adjusted to calendar year 2002, from the 2002 ACS.

Estimates of nonelderly adults who are poor and near-poor are strikingly similar between the CPS and ACS at 45.6 versus 45.3 million low-income persons or 25.8 versus 26.1 percent of the nonelderly adult population (see Table 10-9). The NHIS, however, has substantially more at 53 million or nearly 30 percent of the nonelderly adult population with low income.

The ACS finds the fewest low-income elderly at 11.2 million or 33.3 percent, compared with 13.2 million or 38.5 percent for the CPS (see Table 10-10). Estimates of the number of poor elderly do not differ as much among the three surveys, however, with a range of only 3.2 to 3.8 million or 9.5 to 11.0 percent.

Impact of the Family Definition

Some surveys utilize family definitions that deviate from the CPS family concept, which is incorporated into the official measure of poverty in the United States. The NHIS, as noted, includes unmarried partners and their children in the same family, and it includes foster children as part of the family as well. Broadening the family concept relative to the CPS family produces major changes in family income and poverty rates.

In developing the NHIS estimates of income for comparison with the other surveys, I separated unmarried partners and foster children from

TABLE 10-9 Estimates of Poor and Near-Poor Nonelderly Adults: Three Surveys

Estimate	CPS	ACS	NHIS
	Millions of Persons		
All Adults Aged 18 to 64	176.66	173.34	177.76
Poverty Status			
Poor	18.77	18.91	23.53
Near-poor	26.85	26.36	29.40
Total Low Income	45.62	45.27	52.94
	Percentage of the Population		
All Adults Aged 18 to 64	100.00	100.00	100.00
Poverty Status			
Poor	10.60	10.90	13.20
Near-poor	15.20	15.20	16.50
Total Low Income	25.80	26.10	29.80

NOTES: The poor have a family income below the poverty threshold. The near-poor have a family income at or above the poverty threshold but below twice the poverty threshold.
SOURCES: Mathematica Policy Research, Inc., from tabulations of poverty status in calendar year 2002 from the 2003 CPS ASEC supplement and the 2003 NHIS, and poverty status in the prior 12 months, inflation-adjusted to calendar year 2002, from the 2002 ACS.

TABLE 10-10 Estimates of Poor and Near-Poor Elderly: Three Surveys

Population Subgroup	CPS	ACS	NHIS
	Millions of Persons		
All Persons Aged 65 and Older	34.22	33.56	34.22
Poverty Status			
Poor	3.58	3.20	3.76
Near-poor	9.58	7.98	9.10
Total Low Income	13.16	11.18	12.86
	Percentage of the Population		
All Persons Aged 65 and Older	100.00	100.00	100.00
Poverty Status			
Poor	10.50	9.50	11.00
Near-poor	28.00	23.80	26.50
Total Low Income	38.50	33.30	37.60

NOTES: The poor have a family income below the poverty threshold. The near-poor have a family income at or above the poverty threshold but below twice the poverty threshold.
SOURCES: Mathematica Policy Research, Inc., from tabulations of poverty status in calendar year 2002 from the 2003 CPS ASEC supplement and the 2003 NHIS, and poverty status in the prior 12 months, inflation-adjusted to calendar year 2002, from the 2002 ACS.

TABLE 10-11 Comparison of the CPS and NHIS Family Concepts with Respect to the Estimated Distribution of Persons by Income Relative to Poverty

Family Income as Percentage of Poverty	CPS Family	NHIS Family	Change
Total Percentage	100.00	100.00	
Under 100	14.70	13.70	–0.90
100 to under 200	19.00	19.00	0.00
200 to under 400	30.70	30.90	0.20
400 or more	35.70	36.40	0.80
Total Percentage	283.70	283.90	0.20
Under 100	41.60	39.00	–2.60
100 to under 200	53.90	53.80	–0.10
200 to under 400	87.10	87.70	0.60
400 or more	101.20	103.40	2.30

SOURCE: Mathematica Policy Research, Inc., from tabulations of poverty status in calendar year 2002 from the 2003 NHIS.

the NHIS family and apportioned family income among the two or more family units created from each NHIS family and that conform to the CPS family definition. By comparing the income and poverty estimates prepared using the CPS family definition with estimates obtained from the original data, I assessed the impact of using the NHIS versus CPS family definition on groups and individuals for the purposes of estimating family income.

Table 10-11 shows estimates of the impact of the broader NHIS family definition. The NHIS family definition reduces the number of poor by 2.6 million and reduces the poverty rate by 0.9 percentage points. There is no impact on the percentage of persons between 100 and 200 percent of poverty, which means that the number of people who were moved above the poverty line by the NHIS family concept is offset by the number of people who were moved beyond 200 percent of poverty. Most of the upward shift is observed in the top category—that is, among people above 400 percent of poverty, where the broader family concept adds 2.3 million to the number in the NHIS.

Aspects of Data Quality

Two ways in which respondents can diminish the effectiveness of even very well designed income questions are by providing no answers at all or (which may be worse) by giving inaccurate answers. It is well known that income questions generate some of the highest item nonresponse rates in surveys generally. Frequently, this results in large amounts of missing income data. Unless the data producers choose to leave such missing data for their users to address, they must apply one or more methods of allocation (or imputation) to fill in the missing data. When the data producers elect to allocate their missing income data, high rates of nonresponse are likely to mean that large fractions of the income data they provide to their users will have been created by the data producers themselves rather than supplied by their respondents. This makes the quality of the income data dependent not only on the completeness and accuracy of the reported amounts, but also on the quality of the methods used to generate allocated amounts. High-quality allocations are unbiased, have the appropriate variances, and reflect the covariance structure that would be observed if the missing values were fully and correctly reported.

One can quantify the amount of income data that are allocated in a survey and, in so doing, measure the magnitude of nonresponse and its potential impact on data quality. One cannot assess in any direct way the accuracy of survey responses to income questions. However, one way in which respondents may reduce the accuracy of their responses is to use a high level of approximation—for example, by reporting a salary of $50,000 when the true salary lies somewhere between $45,000 and $55,000. When a significant number of respondents round their responses in this way, it distorts the distribution by creating spikes at the rounded values. In fact, rounding is a commonly used technique for protecting the confidentiality of income data in public-use files.[10] The frequency of rounded responses can be quantified, and I do so for selected income sources for the three surveys.

Nonresponse and Allocation

The percentage of respondents with any income allocated is sensitive to the number of income questions asked in the survey. For this reason, it is more useful to look at the proportion of dollars that was allocated. Table 10-12 reports the fraction of total dollars allocated in the three surveys by source of income. For total income, this fraction in the ACS is about half what it is in the CPS: 17.6 percent versus 34.2 percent. The allocation rate in the NHIS is similar to the CPS at 32.4 percent. For the seven sources of income listed in the table, allocation rates in the CPS vary from a low of 28.0 percent (Supplemental Security Income) to a high of 62.6 percent (asset income). Allocation rates in the ACS show little variation by source,

[10] Limiting the number of significant digits in reported incomes reduces their uniqueness, making them less identifiable. The ACS uses a very well-defined rounding rule.

TABLE 10-12 Percentage of Income Allocated by Source: Three Surveys

Source of Income	CPS	ACS	NHIS
Total Income (NHIS family income)	34.20	17.60	32.40
Wages and salaries (NHIS earnings)	32.00	17.20	31.80
Self-employment	44.70	23.10	n/a
Asset income	62.60	19.40	n/a
Social Security or Railroad Retirement	35.50	18.50	n/a
Supplemental Security Income	28.00	16.70	n/a
Welfare	29.20	17.90	n/a
Pensions	35.40	16.20	n/a

NOTE: n/a = not available.
SOURCES: Mathematica Policy Research, Inc., from tabulations of calendar year 2002 income from the 2003 CPS ASEC supplement; the 2003 NHIS; and prior 12 months income, inflation-adjusted to calendar year 2002, from the 2002 ACS.

however, with the lowest (16.2 percent for pensions) and the highest (23.1 percent for self-employment income) being separated by only 7 percentage points. Thus, the ACS allocation rate for asset income is less than a third of the corresponding rate in the CPS.

One can speculate that the lower nonresponse to the ACS income questions is a carryover from the mandatory nature of the survey. This may also explain why the ACS allocation rates differ so little by source of income. Regardless of the explanation, the markedly lower allocation rate in the ACS is an important indicator of data quality.

Rounding

One reason to examine rounding in the context of policy analytic use of income data is that the heaping of incomes at well-spaced values can distort the results of policy simulations involving the use of income thresholds to establish program eligibility. An eligibility threshold that lies near an income amount with excessive heaping will produce dramatically different results depending on whether the threshold falls just below or just above that amount. If the former, a simulation will mildly understate the impact of a small change in policy; if the latter, it will grossly overstate the impact of the policy change. Another reason to be concerned about rounding is that a high level of rounding suggests inaccuracy or a lack of precision in the reported amounts generally.

I examined the extent of rounding in reported incomes below $52,500 for earnings, wages and salaries, Social Security benefits, other retirement income, total personal income, and total family income in the three surveys.[11] Earnings and total family income are the only income amounts collected in the NHIS and therefore the only amounts on which all three surveys could be compared. I compared the CPS and the ACS with respect to four additional sources. Social Security benefits reported at the person level will have been collected as a single value in both surveys. Most respondents reporting wages and salaries, retirement income, and even earnings are likely to have supplied a single value in response to one question, even though multiple questions were asked.

The results show the differential impact of few versus many income questions. In the NHIS, which relies on a single, person-level question to collect earnings and a single, family-level question to collect total family income, 40 percent of the reported earnings and 36 percent of the reported family incomes below $52,500 are multiples of $5,000, and 23 percent of the earnings and 21 percent of the total family incomes are multiples of $10,000 (see Table 10-13).

On wages and salaries as well as earnings, the CPS is only marginally better than the ACS, with 27 to 28 percent of the amounts being divisible by $5,000 compared with 30 percent for the ACS. The ACS shows markedly more rounding than the CPS on total personal income (20 versus 14 percent) and total family income (16 versus 11 percent). Rounding in both surveys is much lower for Social Security and other retirement income than for the other sources, but rounding in the CPS is still several percentage points lower than in the ACS.

MORE CURRENT COMPARISONS

Comparing just a single year of data provides no basis for assessing how the rolling reference period for annual income data in the ACS may affect the measurement of income and poverty in the ACS relative to the CPS. To provide more data on this issue, I constructed simple comparisons using data from the 2007 and 2008 ACS and the 2008 and 2009 CPS ASEC supplements. In addition, with the 2008 ACS data it is possible to compare the two surveys with respect to their estimates of health insurance coverage by relative income. Below, the first section compares estimates of the poor and near-poor for the total civilian noninstitutionalized population and for children and both nonelderly and elderly adults. The next section compares estimates of the percentage uninsured by income relative to poverty for these same populations. The third section examines the impact of differences between the CPS and the ACS residency rules

[11] I selected $52,500 in order to examine the frequency of rounding up to levels of $50,000. The ACS public-use file has incomes of $50,000 or greater rounded to the nearest $1,000.

TABLE 10-13 Reporting of Rounded Values by Source of Income by Survey Among Positive Dollar Amounts Below $52,500

Income Source and Level of Rounding	CPS	ACS	NHIS
Earnings			
Percentage divisible by $5,000	27.80	29.60	39.80
Percentage divisible by $10,000	15.80	17.40	22.90
Percentage of income in range	82.10	82.40	80.90
Wages and Salaries			
Percentage divisible by $5,000	27.20	29.70	n/a
Percentage divisible by $10,000	15.40	17.40	n/a
Percentage of income in range	82.20	82.70	n/a
Social Security			
Percentage divisible by $5,000	0.60	4.30	n/a
Percentage divisible by $10,000	0.40	1.90	n/a
Percentage of income in range	100.00	100.00	n/a
Retirement Income			
Percentage divisible by $5,000	4.50	8.00	n/a
Percentage divisible by $10,000	2.70	4.30	n/a
Percentage of income in range	95.60	95.40	n/a
Total Personal Income			
Percentage divisible by $5,000	13.70	19.70	n/a
Percentage divisible by $10,000	7.80	11.50	n/a
Percentage of income in range	84.60	84.20	n/a
Total Family Income			
Percentage divisible by $5,000	11.00	16.20	35.60
Percentage divisible by $10,000	6.20	9.50	20.90
Percentage of income in range	66.90	66.00	60.30

NOTES: Allocated amounts are excluded from each source. Family income for the NHIS is based on the NHIS family, which is the level at which family income was reported.
n/a = not available.
SOURCES: Mathematica Policy Research, Inc., from tabulations of calendar year 2002 income from the 2003 CPS ASEC supplement and the 2003 NHIS and prior 12 months income, inflation-adjusted to calendar year 2002, from the 2002 ACS.

by comparing young adults aged 18 to 24 by college enrollment, relative income, and the percentage lacking health insurance coverage.

Poor and Near-Poor

For 2007 the ACS estimated a slightly higher poverty rate than the CPS, at 13.0 versus 12.5 percent, and a slightly smaller fraction between

100 and 200 percent of poverty (near-poor), at 17.7 versus 18.0 percent (see Table 10-14). The total fraction of the population estimated as low income (below 200 percent of poverty) was essentially identical in the two surveys: 30.6 percent in the ACS versus 30.5 percent in the CPS. For 2008, when the United States entered a deep recession in the latter part of the year, the CPS showed an increase of 0.7 percentage points in both the percentage poor and the percentage near-poor, yielding an increase of 1.4 percentage points in the percentage of the population classified as low income. The ACS, with its longer reference period, showed little change between 2007 and 2008, with the poverty rate rising by just 0.2 percentage points and the fraction near-poor remaining essentially unchanged. The fraction of the population classified as low income rose by just 0.3 percentage points in the ACS, leaving the ACS a full percentage point below the CPS. These changes are consistent with the expectation that the ACS should be slower to respond to economic changes.

In effect, the same pattern can be seen among children, with no difference between the two surveys in the percentage with low income in 2007 but the CPS moving more than a full percentage point above the ACS in 2008 (Table 10-15). Among nonelderly adults, the ACS had a percentage point higher fraction classified as low income in 2007 but showed only a 0.4

TABLE 10-14 Estimates of the Poor and Near-Poor: CPS and ACS, 2007 and 2008

	2007		2008	
Estimate	CPS	ACS	CPS	ACS
	Millions of Persons			
All Persons	297.81	293.42	300.10	295.32
Poverty Status				
Poor	37.22	38.01	39.74	39.03
Near-poor	53.64	51.84	56.04	52.33
Total Low Income	90.86	89.85	95.77	91.36
	Percentage of the Population			
All Persons	100.00	100.00	100.00	100.00
Poverty Status				
Poor	12.50	13.00	13.20	13.20
Near-poor	18.00	17.70	18.70	17.70
Total Low Income	30.50	30.60	31.90	30.90

NOTES: The poor have a family income below the poverty threshold. The near-poor have a family income at or above the poverty threshold but below twice the poverty threshold.
SOURCES: Mathematica Policy Research, Inc., from tabulations of poverty status in calendar years 2007 and 2008 from the 2008 and 2009 CPS ASEC supplements and the 2007 and 2008 ACS.

TABLE 10-15 Estimates of the Poor and Near-Poor Children Under 18: CPS and ACS, 2007 and 2008

	2007		2008	
Estimate	CPS	ACS	CPS	ACS
	Millions of Persons			
All Children Under Age 18	73.99	72.66	74.07	72.80
Poverty Status				
Poor	13.32	13.03	14.05	13.23
Near-poor	15.66	15.47	16.00	15.53
Total Low Income	28.98	28.50	30.05	28.76
	Percentage of the Population			
All Children Under Age 18	100.00	100.00	100.00	100.00
Poverty Status				
Poor	18.00	17.90	19.0	18.20
Near-poor	21.20	21.30	21.60	21.30
Total Low Income	39.20	39.20	40.60	39.50

NOTES: The poor have a family income below the poverty threshold. The near-poor have a family income at or above the poverty threshold but below twice the poverty threshold.
SOURCES: Mathematica Policy Research, Inc., from tabulations of poverty status in calendar years 2007 and 2008 from the 2008 and 2009 CPS ASEC supplements and the 2007 and 2008 ACS.

percentage point increase in this fraction between 2007 and 2008, while the CPS fraction increased by 1.6 percentage points (see Table 10-16). Among the elderly, for whom the ACS finds substantially fewer near-poor, the ACS had a 4 percentage point lower fraction with low income in 2007 (see Table 10-17). This difference increased by 0.4 percentage points between 2007 and 2008. The more modest increase among the elderly versus the nonelderly is consistent with the elderly receiving more of their income from sources that are less affected by recession than children and working-age adults.

Uninsured by Relative Income

For the civilian noninstitutionalized population as a whole, the uninsured rate measured in the CPS for 2008 was 15.4 percent. Using a very different approach to measuring health insurance coverage, the ACS obtained an uninsured rate of 15.2 percent (see Table 10-18). The other papers in this session address the differences in measurement and possible reasons why the uninsured rates are nevertheless so close. When persons are classified by income relative to poverty, similar uninsured rates are also found in each of the four classes. The ACS has a somewhat lower rate among the poor (29.1 versus 30.5 percent) but a slightly higher rate

TABLE 10-16 Estimates of the Poor and Near-Poor Nonelderly Adults: CPS and ACS, 2007 and 2008

	2007		2008	
Estimate	CPS	ACS	CPS	ACS
	Millions of Persons			
All Adults Aged 18 to 64	187.03	184.53	188.25	185.32
Poverty Status				
Poor	20.35	21.52	22.03	22.12
Near-poor	28.28	28.23	30.00	28.63
Total Low Income	48.63	49.75	52.03	50.75
	Percentage of the Population			
All Adults Aged 18 to 64	100.00	100.00	100.00	100.00
Poverty Status				
Poor	10.90	11.70	11.70	11.90
Near-poor	15.10	15.30	15.90	15.50
Total Low Income	26.00	27.0	27.60	27.40

NOTE: The poor have a family income below the poverty threshold. The near-poor have a family income at or above the poverty threshold but below twice the poverty threshold.
SOURCES: Mathematica Policy Research, Inc., from tabulations of poverty status in calendar years 2007 and 2008 from the 2008 and 2009 CPS ASEC supplements and the 2007 and 2008 ACS.

among the near-poor (25.9 versus 24.3 percent). For the combined, low-income population, the uninsured rates for the two surveys are therefore quite similar: 26.9 percent for the CPS and 27.3 percent for the ACS. The rates are essentially identical among persons between 200 and 400 percent of poverty, whereas the ACS rate is modestly lower among persons above 400 percent of poverty: 5.5 percent versus 6.0 percent in the CPS.

Among children, the ACS uninsured rate is fractionally higher than the CPS at 9.9 versus 9.8 percent (see Table 10-19). Uninsured rates are equally similar among poor children, whereas the ACS uninsured rate is more than a percentage point higher among the near-poor: 15.9 versus 14.7 percent. As with the population as a whole, the ACS is 0.1 percentage point higher among children between 200 and 400 percent of poverty and 0.5 percentage points lower among children above 400 percent of poverty.

The pattern changes for nonelderly adults, for whom the poor in the ACS are somewhat less likely to be uninsured than the poor in the CPS: 41.3 versus 44.0 percent (see Table 10-20). The ACS is a percentage point higher among the near-poor, and the differences among those above 200 percent of poverty are identical to those for children, even though the uninsured rates for nonelderly adults are more than twice as high as they are for children.

TABLE 10-17 Estimates of the Poor and Near-Poor Elderly: CPS and ACS, 2007 and 2008

	2007		2008	
Estimate	CPS	ACS	CPS	ACS
	Millions of Persons			
All Persons Aged 65 and Older	36.79	36.23	37.79	37.20
Poverty Status				
Poor	3.56	3.46	3.66	3.68
Near-poor	9.70	8.14	10.03	8.16
Total Low Income	13.26	11.59	13.69	11.84
	Percentage of the Population			
All Persons Aged 65 and Older	100.00	100.00	100.00	100.00
Poverty Status				
Poor	9.70	9.50	9.70	9.90
Near-poor	26.40	22.50	26.60	21.90
Total Low Income	36.00	32.00	36.20	31.80

NOTE: The poor have a family income below the poverty threshold. The near-poor have a family income at or above the poverty threshold but below twice the poverty threshold.
SOURCES: Mathematica Policy Research, Inc., from tabulations of poverty status in calendar years 2007 and 2008 from the 2008 and 2009 CPS ASEC supplements and the 2007 and 2008 ACS.

TABLE 10-18 Number and Percentage Uninsured by Relative Income: All Persons, CPS and ACS, 2008

Estimate	CPS	ACS
	Uninsured Persons (millions)	
All Persons	46.25	44.76
Income Relative to Poverty		
Under 100%	12.12	11.37
100 to under 200%	13.64	13.54
200 to under 400%	13.80	13.63
400 or more	6.68	6.22
Low Income (under 200%)	25.76	24.91
	Percentage Uninsured	
All Persons	15.40	15.20
Income Relative to Poverty		
Under 100%	30.50	29.10
100 to under 200%	24.30	25.90
200 to under 400%	14.80	14.90
400 or more	6.00	5.50
Low Income (under 200%)	26.90	27.30

SOURCE: Mathematica Policy Research, Inc., from tabulations of poverty status in calendar year 2002 from the 2003 NHIS.

TABLE 10-19 Number and Percentage Uninsured by Relative Income: Children Under 18, CPS and ACS, 2008

Estimate	CPS	ACS
	Uninsured Persons (millions)	
All Children Under Age 18	7.25	7.20
Income Relative to Poverty		
Under 100%	2.21	2.10
100 to under 200%	2.35	2.48
200 to under 400%	1.99	2.01
400 or more	0.70	0.61
Low Income (under 200%)	4.56	4.58
	Percentage Uninsured	
All Children Under Age 18	9.80	9.90
Income Relative to Poverty		
Under 100%	15.70	15.90
100 to under 200%	14.70	15.90
200 to under 400%	8.60	8.70
400 or more	3.40	2.90
Low Income (under 200%)	15.20	15.90

SOURCE: Mathematica Policy Research, Inc., tabulations of the 2009 CPS ASEC supplement and the 2008 ACS.

Uninsured rates among the elderly are not given much attention in the literature, as they are exceedingly low due to the broad coverage provided by Medicare, and the estimates are probably dominated by measurement error. Compared with the CPS, the ACS finds a markedly lower uninsured rate among the poor elderly—3.6 versus 6.2 percent—but at higher levels of relative income, the estimates are remarkably similar (see Table 10-21).

Residency Among Young Adults

By asking that college students living away from home be counted at their place of residence at the time of the survey, the ACS departs from the CPS in a way that could lead to higher estimates of the college student population. College students living away from home are given two opportunities to be counted, in effect: once at school and once at home. While the ACS asks that students be counted at school, parents responding to the survey may choose to count their children at home regardless. The fact that the ACS includes college dormitories in its sample frame while the CPS does not means that every college student has a chance to be counted twice in the ACS, whereas students in dormitories are precluded from this possibility in the CPS.

TABLE 10-20 Number and Percentage Uninsured by Relative Income: Nonelderly Adults, CPS and ACS, 2008

Estimate	CPS	ACS
	Uninsured Persons (millions)	
All Nonelderly Adults	38.35	37.03
Income Relative to Poverty		
Under 100%	9.68	9.14
100 to under 200%	11.12	10.91
200 to under 400%	11.66	11.47
400 or more	5.88	5.50
Low Income (under 200%)	20.80	20.05
	Percentage Uninsured	
All Nonelderly Adults	20.40	20.00
Income Relative to Poverty		
Under 100%	44.00	41.30
100 to under 200%	37.10	38.10
200 to under 400%	20.30	20.40
400 or more	7.50	7.00
Low Income (under 200%)	40.00	39.50

SOURCE: Mathematica Policy Research, Inc., from tabulations of the 2009 CPS ASEC supplement and the 2008 ACS.

Comparative estimates from the two surveys for 2008 show 1.8 million more young adults aged 18-24 enrolled in college in the ACS than the CPS; 1.0 million fewer young adults not enrolled in college in the ACS than the CPS; and almost 0.9 million more young adults in total in the ACS than the CPS (see Table 10-22). Despite students in dormitories being excluded from the ACS poverty universe, it still finds a higher fraction of young adults to be poor (20.0 versus 18.4 percent). Virtually all of the difference is due to college students, whose poverty rate is 5 percentage points higher in the ACS than the CPS despite the exclusion of 18.7 percent of students from the ACS poverty universe. Among those who are not enrolled in college, virtually none is excluded from the ACS poverty universe, and the distribution of persons across the four classes of relative income is quite similar between the ACS and the CPS. It is clear from the comparative distributions that the students excluded from the poverty universe in the ACS are drawn from higher income levels, as that is where the ACS falls well short of the CPS.

A comparison of uninsured rates by college enrollment and relative income provides clear evidence that the college students identified as poor in the ACS look more like students from higher income families in the CPS. The uninsured rate of 19.9 percent for poor college students in the ACS compares with 32.5 percent in the CPS, whereas both surveys

TABLE 10-21 Number and Percentage Uninsured by Relative Income: Elderly Persons, CPS and ACS, 2008

Estimate	CPS	ACS
	Uninsured Persons (millions)	
All Persons Aged 65 and Older	0.65	0.54
Income Relative to Poverty		
Under 100%	0.22	0.13
100 to under 200%	0.17	0.15
200 to under 400%	0.15	0.15
400 or more	0.10	0.11
Low Income (under 200%)	0.39	0.28
	Percentage Uninsured	
All Persons Aged 65 and Older	1.70	1.40
Income Relative to Poverty		
Under 100%	6.20	3.60
100 to under 200%	1.70	1.80
200 to under 400%	1.20	1.20
400 or more	0.90	0.80
Low Income (under 200%)	2.90	2.40

SOURCE: Mathematica Policy Research, Inc., from tabulations of the 2009 CPS ASEC supplement and the 2008 ACS.

find uninsured rates around 20 percent for college students with incomes between 200 and 400 percent of poverty (see Table 10-23). Among those not enrolled in college, the ACS has somewhat higher uninsured rates in every category of relative income. I have no explanation for this. Regardless of college enrollment, only 6 to 7 percent of young adults excluded from the ACS poverty universe are reported to be without health insurance coverage.

IMPLICATIONS

The principal findings can be summarized quite simply. At high levels of aggregation, the ACS produces estimates of income and health insurance coverage that look strikingly similar to those obtained from the CPS, despite notable differences in measurement. This suggests but by no means proves that, with its much larger sample size, the ACS could be used to develop direct estimates of low-income uninsured children at the state level and even lower levels of geography that would mimic what could be obtained from the CPS if it had a much bigger sample.

Although the ACS seems to fare well with a modest set of income questions, the NHIS illustrates potential adverse consequences from devoting

TABLE 10-22 Persons Aged 18 to 24 by College Enrollment and Relative Income: CPS and ACS, 2008

Estimate	CPS	ACS
	Millions of Persons	
All Persons Aged 18 to 24	28.47	29.31
	Percentage of Persons	
Income Relative to Poverty	100.00	100.00
Outside the poverty universe	0.00	7.90
Under 100%	18.40	20.00
100 to under 200%	20.40	18.90
200 to under 400%	31.10	27.80
400% or more	30.00	25.40
	Millions of Persons	
Enrolled in College	10.40	12.22
	Percentage of Persons	
Income Relative to Poverty	100.00	100.00
Outside the poverty universe	0.00	18.70
Under 100%	15.60	20.70
100 to under 200%	16.70	12.90
200 to under 400%	28.10	21.10
400% or more	39.60	26.60
	Millions of Persons	
Not Enrolled in College	18.07	17.08
	Percentage of Persons	
Income Relative to Poverty	100.00	100.00
Outside the poverty universe	0.00	0.20
Under 100%	20.10	19.50
100 to under 200%	22.50	23.20
200 to under 400%	32.90	32.60
400% or more	24.50	24.60

SOURCE: Mathematica Policy Research, Inc., tabulations of the 2009 CPS ASEC supplement

too few questions to income measurement—particularly among families at the low end of the income distribution. Also, differences between the ACS and the CPS begin to emerge as income and/or the population are disaggregated. Such differences are often but certainly not always consistent with differences in the two surveys' approaches to measuring income, including reference period as well as the way that income is defined. For

TABLE 10-23 Percentage Uninsured among Persons Aged 18 to 24 by College Enrollment and Relative Income: CPS and ACS, 2008

Estimate	CPS	ACS
All Persons Aged 18 to 24	28.80	28.60
Income Relative to Poverty		
Outside the poverty universe	n/a	6.30
Under 100%	41.90	36.60
100 to under 200%	39.80	43.00
200 to under 400%	28.70	30.90
400% or more	13.40	15.90
Enrolled in College	19.10	15.90
Income Relative to Poverty		
Outside the poverty universe	n/a	6.30
Under 100%	32.50	19.90
100 to under 200%	30.60	29.60
200 to under 400%	20.40	20.70
400% or more	8.10	9.00
Not Enrolled in College	34.40	37.70
Income Relative to Poverty		
Outside the poverty universe	n/a	6.70
Under 100%	46.10	49.40
100 to under 200%	43.80	48.30
200 to under 400%	32.80	35.70
400% or more	18.30	21.20

NOTE: n/a = not available.
SOURCE: Mathematica Policy Research, Inc., from tabulations of the 2009 CPS ASEC supplement and the 2008 ACS.

example, in recent years the ACS lags the CPS in its response to changes in the economy—as it should, given the longer reference period in the ACS. This could be viewed as a drawback for many prospective uses of the ACS, but for state- and lower level estimates it is important to remember that the Census Bureau combines 3 years of CPS data to produce the annual estimates of low-income uninsured children that the ACS would potentially replace. Compared with a 3-year moving average, the ACS data show greater rather than lesser sensitivity to economic change.

The Census Bureau and the research community are only beginning to understand the implications of many aspects of ACS data collection. From the perspective of income measurement, the chief issues revolve around the ACS reference period and the ways that the Census Bureau has attempted to annualize the estimates, the residence rules for college students living away from home, the small number of questions used to capture the diverse sources of unearned income, and the absence of data on the relationships among persons who are unrelated to the householder.

ACKNOWLEDGMENTS

Much of the work presented herein was prepared by Mathematica Policy Research, Inc., under contract to the Office of the Assistant Secretary for Planning and Evaluation, U.S. Department of Health and Human Services and presented in a final report, "Income Data for Policy Analysis: A Comparative Assessment of Eight Surveys" dated December 23, 2008, and coauthored by Gabrielle Denmead. The findings and conclusions reported here are those of the author and do not necessarily represent the views of ASPE or HHS.

REFERENCE

Czajka, J.L., and Denmead, G. (2008). *Income Data for Policy Analysis: A Comparative Assessment of Eight Surveys*. Report from Mathematica Policy Research, Inc. Available: http://www.mathematica-mpr.com/publications/pdfs/incomedata.pdf [October 2010].

11

Using Uninsured Data to Track State CHIP Programs

John McInerney
The Commonwealth Institute for Fiscal Analysis

With the creation of the Children's Health Insurance Program (CHIP) as part of the Balanced Budget Act of 1997, focus and attention on providing access to health insurance coverage for low-income children significantly increased. CHIP gave states an attractive option to help reduce the population of low-income uninsured children, which was in excess of 20 percent in the years leading up to the program's enactment. Combined with a generous federal contribution, the block grant program helped states provide comprehensive and affordable coverage to children with family income too high for Medicaid but too low to afford private insurance coverage.[1]

Throughout the program's history, data on the uninsured rate of children has played an important role in funding and the evaluation of program success. Nationwide, as the number of low-income children enrolled in CHIP programs has increased, the uninsured rate for children has fallen according to census data and individual state surveys. CHIP and Medicaid became successful examples of using public insurance options effectively to make real progress in reducing the rates of uninsured. The recently enacted Patient Protection and Affordable Care Act (PPACA) builds on that success. When implemented over the next several years,

[1] Although CHIP's income eligibility maximum was set at 200 percent of the federal poverty level, many states have used income disregards to expand eligibility to higher income children. In addition, states with higher Medicaid eligibility were able to set their CHIP eligibility at 50 percentage points above the Medicaid endpoint.

the law will retain CHIP and Medicaid as a primary source of coverage for low-income children and families.

State policy makers, advocates, and child coverage experts have learned a great deal about the use of data in evaluating CHIP over time. Using Virginia as a primary example, this paper discusses the some of these lessons in using uninsured and administrative data in running and evaluating CHIP at the state level. First, I discuss the role that data have played in CHIP policy making and program management and also how state analysts and advocates use, promote, and disseminate national survey data at the state level. I then describe how more accessible local uninsured and enrollment data would add useful information to aid in evaluating CHIP and the unmet needs of uninsured children.

THE VIRGINIA CHIP EXPERIENCE

Virginia, like most states, had high levels of uninsurance among children before CHIP was created in 1997. The percentage of employer-sponsored health insurance (ESI) for children was in decline, and there were few public insurance options for children whose families had too much income to qualify for Medicaid coverage. Nationally, over 22 percent of low-income children with family incomes below 200 percent of the federal poverty line (FPL) were without health insurance in 1997 (Georgetown University Center for Children and Families, 2006).

The Balanced Budget Act allocated $40 billion over 10 years in federal funding to allow states to create CHIP programs. Unlike Medicaid, CHIP was created as a block grant program, so states would receive an allotment based on a funding formula, not on the number of eligible applicants. States were given significant flexibility in program design and management in determining eligibility levels, enrollment, and program design. To further entice states, the federal match rate was set at 30 percent above the match rate for Medicaid. A state with a 50 percent federal match rate in Medicaid would receive 65 percent federal reimbursement for CHIP.[2] States broadly embraced CHIP, and, by 2000, every state had a program up and running.

However, the effort was not the same in all states. Despite the substantial uninsured low-income child population, Virginia did not fully embrace CHIP initially. The state program, the Children's Medical Security Insurance Plan (CMSIP), was created as a Medicaid expansion program, covering children from the Medicaid eligibility endpoint[3] up to 185

[2] Virginia's CHIP match rate is 65 percent.

[3] Medicaid eligibility for children aged 6-19 was 100 percent of FPL; children under age 6 were covered up to 133 percent of FPL.

percent of FPL. Enrollment in the program was low, largely because of enrollment barriers, such as a 12-month waiting period for children who had other insurance coverage.

The lackluster results of CMSIP were seen as a failure to many in the state, and the program was reformed in August 2001. The name of the program was changed to Family Access to Medical Insurance Security (FAMIS), eligibility was increased to 200 percent of FPL, and the 12-month waiting period was shortened to 6 months. In addition, the program was changed from a Medicaid expansion program to a separate CHIP. Further reforms in 2002 and 2003 implemented a common application for FAMIS and Medicaid, created a 12-month continuous eligibility provision for FAMIS children, and streamlined Medicaid eligibility for children at 133 percent of FPL (with children aged 6-18 and family incomes between 100-133 percent of FPL covered as a Medicaid expansion population with CHIP funding).

Since the 2001-2003 reforms, FAMIS has remained relatively unchanged.[4] Although the state was utilizing all of its federal funding, Virginia was able to use a significant carryover balance from the slow beginnings of CHIP coverage to avoid the kind of funding shortfall that many other states faced. The temporary extension of CHIP in 2007 and its reauthorization in February 2009 also aided the Commonwealth in avoiding an estimated $24 million shortfall beginning in July 2009. The Children's Health Insurance Program Reauthorization Act (CHIPRA) will provide states with a total of $69 million over 5 years in federal funding. The new CHIP block grant under CHIPRA provides Virginia with $175 million and $188 million in federal funding for the first 2 years of the 5-year reauthorization and is sufficient to provide coverage at the current enrollment and eligibility levels.

ADMINISTRATIVE DATA SHOW PROGRAM ENROLLMENT INCREASES BUT CURRENT POPULATION SURVEY DATA SHOW RISE IN UNINSURED

Virginia's CHIP enrollment dramatically increased in the years following the creation of FAMIS. In 2001, fewer than 38,000 were enrolled

[4] One change was adding coverage for low-income pregnant women through the FAMIS MOMS program beginning in 2005 for women with family income up to 150 percent of poverty (beginning when Medicaid coverage ends at 133 percent of FPL). The program has been expanded and now covers women up to 200 percent of FPL. In addition, the Medicaid citizenship documentation requirement of the 2005 Deficit Reduction Act caused many children to lose Medicaid coverage in Virginia. The state saw a net decrease of over 13,000 in the first 8 months. However, FAMIS was not affected by the Medicaid citizenship documentation rule, and enrollment increased by over 2,500 during the same period.

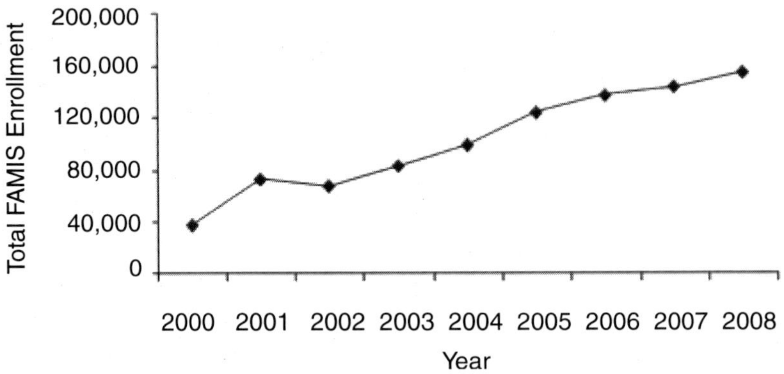

FIGURE 11-1 CHIP coverage for kids steadily increased in Virginia, 2000 to 2008.
SOURCE: Centers for Medicare & Medicaid Services (CMS).

in the program, far lower than most other comparable states. However, enrollment steadily increased each year, and, by 2006, approximately 125,000 were enrolled at some point during the year (see Figure 11-1). Virginia's outreach was extensive, and the program proved popular with the public. The Commonwealth even renamed its Medicaid program for children FAMIS Plus to capitalize on the good public support.

Yet despite enrollment gains in FAMIS and Medicaid, data from the Current Population Survey (CPS) suggested that progress in covering uninsured children in Virginia was inconsistent. Between 2000 and 2004, CPS data showed a 5 percentage point decline in the low-income uninsured population, which was 23.5 percent in 2000 (this decline, however, was not statistically significant). This decline corresponded with the introduction of CHIP in Virginia and expansions in the program, and these trends are consistent with CHIP lowering the rate of uninsurance for Virginia's children. However, between 2004 and 2006—a period during which FAMIS enrollment increased steadily—CPS data show that the percentage of low-income uninsured children in Virginia actually increased by nearly 4 percentage points (this increase, however, was not statistically significant) (Cook, Kenney, and Lawton, 2010).[5]

Data users are left to provide a plausible explanation for why the CPS shows that Virginia's uninsured rate increased during this period

[5] The Census Bureau revised the assignment of dependent coverage, which impacted the 2004 estimates. Following Cook, Kenney, and Lawton (2010), I use the unrevised 2004 data when discussing changes between 2000 and 2004 (the unrevised uninsurance rate for 2004 is 18.5 percent), and the revised data to discuss changes between 2004 and later years (the revised uninsurance rate for 2004 is 15.9 percent).

of expansion in CHIP.[6] It is likely that an erosion in ESI coverage for all incomes led to the increase in uninsured children. Traditionally, Virginia has had a high percentage of residents with ESI coverage in part because of its large federal workforce. But between 2004 and 2006, ESI coverage in Virginia declined sharply for all incomes (Cook, Kenney, and Lawton, 2010.

LESSONS FOR STATE POLICY MAKERS, ANALYSTS, AND ADVOCATES

Although, in Virginia, CPS data do not reflect the CHIP enrollment gains evident in administrative data between 2004 and 2006, in most states survey and administrative data have shown a negative relationship between CHIP enrollment gains and the uninsured rate for children as measured by survey data. These trends in the data helped encourage state policy makers to do more to increase coverage for children throughout the first decade of CHIP. The results could be seen just about every time the Census Bureau released new data on uninsured rates. While ESI coverage was beginning to erode in many states, and the uninsured rate of the adult nonelderly population was generally stagnant or increasing, state policy makers could point to the positive data on children as a bright spot.[7]

In examining the ways in which data have been used in evaluating and operating CHIP, several lessons learned can help policy makers and those working on children's health issues as the program continues over the next several years.

[6] Nationwide, the uninsured rate for children had also increased during this time period and CHIP enrollment had begun to slow; however, Virginia had actually seen a consistent increase in CHIP enrollment during this period, whereas nationwide enrollment was roughly flat.

[7] However, as the money in the block grant began to be insufficient and many states reached or neared a shortfall in funding, enrollment gains began to slow. Enrollment reached 6 million in federal fiscal 2003. Yet, by FY 2005, it had increased only to 6.2 million. The uncertainty of reauthorization in 2007, ultimately resulting in a temporary extension of the program with increased funding, also hurt the momentum of the program. Correspondingly, the uninsured rate for children remained the same or increased in many states in 2007 and 2008. The reauthorization delay was coupled with a downturn in the economy, forcing a number of states to forgo expansion plans or even, in a few cases, contemplate enrollment freezes or eligibility reductions.

Lesson 1—Uninsured Data for Children and Administrative Enrollment CHIP and Medicaid Data Are Important in Shaping and Directing Policy Options and Decisions

Uninsured Data Used in Original CHIP Funding Formula

Before CHIPRA was enacted in 2009, the CHIP funding formula was a combination of the health costs in the state, the total number of low-income children in the state, and the number of low-income uninsured children in the state.

Because CHIP is a block grant—not an entitlement program like Medicaid—states relied on adequate allotments to meet the demand for coverage. Few states wanted to close the doors to their program or deny eligible children, especially with the generous match, which can go as high as 85 percent.[8] The fact that a part of the state funding allotments was based on CPS uninsured estimates increased the relative importance of the survey in policy making.

However, the importance of the CPS was problematic for many states. Many did not believe that the survey accurately reflected the number of low-income uninsured children in their states. For 42 states the CPS low-income children sample size was less than 100, which often led to funding fluctuations from year to year of up to 25 percent (Blewett and Davern, 2007). Some states received higher allotments than they needed and had large carryover balances and unspent block grant funds. Others were spending their entire block grant and using carryover money from the early years of the program to avoid a funding shortfall. In addition, states were in effect penalized as they covered more children because they were reducing the percentage of low-income children. By the end of 2006, however, at least 14 states faced shortfalls that cumulatively totaled more than $700 million (Park and Broaddus, 2006).

Policy makers recognized these concerns, and CHIPRA created a new funding structure that removes uninsured data from the formula. Instead, states are funded on the basis of the higher of prior year spending or estimated current year spending. In 2011, funding will be rebased so that states will receive funds based on their spending history and need. Survey data can be a helpful indicator in program evaluation, but they were not always an efficient metric in determining CHIP funding allocations.

---

[8] Federal match rates for CHIP run between 65-85 percent. Mississippi has the highest current match rate in FY 2010 at 83 percent.

August 2007: Regulating Income Eligibility Using New CMS Data Requirement

Data have also been used in policy making as a way to control state flexibility in setting income eligibility standards. In August 2007, while Congress was considering reauthorization of CHIP, the Centers for Medicare & Medicaid Services (CMS) issued a directive to prevent states from covering children above 250 percent of FPL unless they met certain requirements. Among the provisions in what became known as the "August 17 directive" was a requirement that states prove that they are covering 95 percent of the eligible children below 200 percent of FPL before offering CHIP coverage to anyone with family income above 250 percent of FPL.

The directive was troubling to states for many reasons. The arguably arbitrary benchmarks, which also included a 12-month waiting period requirement, would have prevented most states from expanding and improving their programs. In addition, states that were already covering children above 250 percent of FPL would have been forced to scale back eligibility if they could not reach the requirement for coverage of uninsured children under 200 percent of FPL. A likely intent of the directive was to set CHIP coverage limits at a maximum of no more than 250 percent of the FPL.

A key concern of state policy makers and other children's health advocates and analysts was the lack of a reliable and uniform data source to measure whether a particular state had actually covered 95 percent of eligible children in CHIP and/or Medicaid.[9]

CMS did not provide states with a definitive data source or method to determine compliance. The CPS would have been the logical choice, but its small sample size and tendency to underreport Medicaid and CHIP coverage concerned the states. Other surveys, such as the Survey of Income and Program Participation and the National Health Interview Survey, did not have recent enough data. Finally, although CMS said that states could use state-based surveys, many were not produced consistently enough to be a viable option (National Academy for State Health Policy, 2008).

In the end, the directive was formally rescinded in February 2009, and no state was forced to reduce income eligibility levels that were already in place. However, CMS did use the directive to initially deny expansion efforts in New York, Ohio, and other states. Essentially, state CHIP coverage was frozen in place for at least a year through the imposition of an

---

[9] Furthermore, it was unlikely that all but a few states would be able to reach the 95 percent requirement. Enrollment rates for even universal programs like Medicare Part A struggle to achieve 95 percent participation rates.

arbitrary set of rules without a viable method of compliance. In the future, as program requirements incorporate data measures, CMS or Congress should be sure to name (and provide) a data source and methodology to determine compliance.

Lesson 2—State Analysts and Advocates Are Key in Disseminating Information to Policy Makers, the Media, and the Public

State groups often lack the resources and capacity to collect and produce survey data, instead relying on national state-specific surveys by the Census Bureau and comprehensive analysis of census and other uninsured data by national experts. State analysts fill a valuable role in explaining the data and promoting state-specific information that can help inform the policy process.

Census Bureau Release of Uninsured Data

One of the most effective ways for children's coverage experts to influence policy makers and the public is through the media. In the current information age, the media includes not only newspapers and television, but also blogs, Twitter, and even social networking sites like Facebook. To capture the attention of these media, old news about the uninsured is a tough sell. Media outlets respond to news that adds fresh information to the discussion ("Are things getting better or worse?" "What does the latest data tell us about where we stand?").

To cultivate the media, one of the most important times of the year for health and social policy analysts and advocates are the days that the Census Bureau releases data from the CPS and the American Community Survey (ACS) on the number and percentage of uninsured people. These are one of the few times of the year that the media are focused and fully engaged on how states and the country are faring in tackling the problem of uninsurance.

Throughout much of the past decade, the message in the CPS release was often clear and easy to articulate to the media, interested policy makers, and the public. The number and percentage of uninsured Americans tended to increase, and the percentage of children without insurance usually fell. CHIP and Medicaid were providing coverage to millions more children. The message was that, without these sources of public insurance, the uninsured rate would undoubtedly be higher.

The difference between a rising uninsured rate overall and a falling uninsured rate for children was a powerful incentive for many state leaders to keep moving ahead with improvements to CHIP through coverage eligibility expansions, enrollment simplifications, and other outreach strategies. Governors and legislators could highlight the success they were having in lowering the uninsured rate for children, diverting attention away from a rising overall uninsured rate, for which they did not have an easy solution at the state level.

Survey Estimates of State Data Fluctuate Year to Year

As mentioned earlier, a common complaint about the CPS state estimates is that the small sample size can sometimes produce inconsistent results or outliers. Most accept this reality, believing that year-to-year results will at least show similar trends, albeit with different levels of magnitude, to larger surveys. Analysts can still present the data in comparison to previous years and help evaluate progress.

However, sometimes trends in survey data can be difficult to explain to a nontechnical audience. As an example, in Virginia the CPS estimated that the state had 185,000 (10.1 percent) uninsured children in 2006 and 187,000 (10.2 percent) in 2007, showing relative stability. Enrollment in FAMIS had increased during these years, but not enough to offset a significant decline in ESI coverage for children.

When the data for 2008 were released in September 2009, most experts expected to see an increase, or at least no significant change, in children's uninsured rates in Virginia. The economy had been in recession for a year, and unemployment had increased from 3.2 percent in December 2007 to 5.2 percent in December 2008.[10] Virginia's overall economy was performing better than many states, but there were still struggles. Yet despite the downturn, the CPS showed Virginia with only 129,000 uninsured children in 2008, a decline of over 30 percent from the previous year. Even using 2- and 3-year averages to help smooth out irregularities and the margin of error in the data, the percentage of uninsured children declined between 2007 and 2008.

Although the uninsured rate for children declined nationally as well, the decline was largely caused by an increase in Medicaid and CHIP enrollment due to the economic downturn. Virginia's decline could not be as easily explained. Public insurance coverage in Virginia increased, according to Census Bureau estimates, but only by less than 1 percentage point. In Virginia, the CPS data suggested that the biggest reason for the decline in the uninsured rate was the increase in ESI coverage for children, which rose from 61.1 percent in 2007 to 69.2 percent in 2008, even though a major recession was under way.

[10] U.S. Department of Labor, Bureau of Labor Statistics: Local Area Unemployment Statistics.

Although analysts could discern that the decline in uninsured children seen in the 2008 CPS data could be explained by an increase in children covered by ESI in the 2008 CPS, these trends in the survey data did not appear to reflect the reality of the situation in the state. It made the job of explaining the results to policy makers and the media more difficult (especially in states that are not so committed to supporting CHIP and Medicaid). Many wondered how it could be that advocates were asking for FAMIS eligibility to be increased when it appeared that the state was achieving success in reducing the uninsured.

Since there can be such large variation in annual estimates, analysts and advocates are left with a difficult decision about how to publicize estimates. They could have a difficult time explaining a good result, such as Virginia's dramatic decline in the number of uninsured children in 2008, as an outlier or an artifact of small sample sizes in a survey. It may be more appropriate to simply point out the unmet needs that still exist (over 100,000 Virginia children may still be eligible but unenrolled for FAMIS and Medicaid) (Holahan, Cook, and Dubay, 2007). If the result is indeed an outlier, then the 2009 data may produce a result more consistent with the overall economic conditions and trends in the state.

Lesson 3—More Data Are Needed

State Surveys Help Fill Gaps, But Are Not Done Regularly in Each State

Although the CPS and the ACS state data are not without flaws, they are the only options for many states in examining uninsured data for children. Most states have produced at least a few state surveys of the uninsured over the past decade, but only a few produce them on a yearly basis. The states that do state-based surveys more often tend to be ones that have a more robust public safety net and lower rates of uninsurance. Surveys, unless funded from federal grants, cost states money, and it is logical that the ones that are more dedicated to reducing the uninsured are more likely to make the investment.

Virginia produced two state surveys in the past decade, the most recent one in 2004.[11] Without federal funding, it is unlikely that the Commonwealth will do another one in the near future. There is no groundswell of support among lawmakers, regardless of ideology, to expand efforts to reduce the uninsured population. In fact, in the most recent bud-

[11] "2004 Virginia Health Care Insurance and Access Survey," produced by the State Health Access Data Assistance Center for the Virginia Department of Health. Funded by the Health Resources and Services Administration State Planning Grants Program.

get proposals, the General Assembly and the governor actively tried to decrease FAMIS eligibility from 200 to 185 percent of FPL to save money in balancing the budget. Although that cut will not happen because of PPACA maintenance-of-effort requirements,[12] Virginia is likely to resist any coverage expansion beyond what is required by the law. In this climate, most policy makers simply do not believe that state surveys are necessary to produce something above and beyond what the Census Bureau produces.

For states that do not do comprehensive surveys, there are ways to get a better picture of state insurance trends. For example, every 2 years, the Virginia Health Care Foundation funds a report by the Urban Institute examining the CPS data and looking at uninsured rates using various demographics like age, gender, and income. The Urban Institute compares the results in Virginia with survey data dating back to 2000. Consumers of the report can see how much progress the state has made in covering children and other populations. For example, has the racial disparity in Virginia improved or deteriorated? Are there more or fewer uninsured now than 8 years ago? Four years ago? Although the data are not new, the Urban Institute study presents a more complete picture that can be used by those seeking to improve and expand CHIP and other children's coverage options.

County Data Would Be Very Valuable

State-level data are not always sufficient to identify where the uninsured are located. Many states, especially larger ones with different regional economic conditions, have areas of both low and high uninsurance. Virginia provides an instructive example. Northern Virginia, near Washington, DC, is one of the most prosperous metropolitan areas in the country, with high incomes and low uninsured rates, in part because of the large federal workforce. Yet by contrast, Southwest Virginia in the Appalachian region contains some of the poorest communities in the United States. Providing one-size-fits-all state data is not representative of the entire makeup of a bigger state.

The Census Bureau is moving toward providing more local data. The Small Area Health Insurance Estimates (SAHIE) Program, with 2006 data most recently released, provides county-level estimates of the uninsured using CPS data. Subsequent years of SAHIE data (the 2007 data are sched-

[12] Virginia did approve, in March 2010 as part of the FY 2011-2012 biennium budget, a reduction in eligibility in FAMIS from 200 to 185 percent of FPL, scheduled to begin in July 2011. However, they will have to reverse the reduction during the next legislative session.

uled for release later in 2010) will also be useful in identifying trends and areas of states that need extra attention.

In addition, the ACS will soon be a significant source of county and regional data for states once the survey has produced 3 years of data. By 2011, the ACS will release 3-year estimates of the uninsured by state, congressional district, metropolitan statistical areas (MSAs), public-use microdata areas (PUMAs), and county. Current 1-year estimates are already available for all these geographic breakdowns except the county level.

Getting access to county information would be beneficial to anyone analyzing CHIP or insurance coverage in general. MSAs and PUMAs provide valuable and important data, but there are limitations. Different municipalities could have vastly different economic conditions, even though they are in the same PUMA or MSA. For example, Virginia Beach is the largest and one of the wealthier cities in the state, and it is in the same MSA as the bordering city of Chesapeake, a much poorer city with a considerably higher uninsured rate and low-income population. Thus, the Norfolk–Virginia Beach–Newport News MSA results do not present an entirely accurate story for either the wealthier or poorer communities. PUMAs are somewhat smaller in size than MSAs, but they face the same problems in providing an accurate assessment when different economic conditions exist within the region.

Because of this problem, local service providers and foundations are not very interested in the MSA and PUMA data if they do not accurately reflect their service areas. They would like local data to help assess their efforts in reaching residents in need in their communities. State-level estimates of the uninsured or CHIP are not particularly relevant either, and regional data could be distorted depending on the breakdown of the communities in the survey.

Privacy and Administrative Data

A further difficulty also exists in analyzing CHIP and Medicaid enrollment at the local level. While some states release these data by county, Virginia will not publicly release the data for most counties. The state interprets the Health Insurance Portability and Accountability Act privacy rules to prohibit the release of enrollment data for jurisdictions with fewer than 20,000 enrollees. This means that data can be released only for the biggest counties or with counties grouped into regional data. With FAMIS, a much smaller program than Medicaid, county-level administrative CHIP data are also not released in Virginia.

This presents a challenge in analyzing the success of CHIP and Medicaid in particular regions of a state. Although state-level administrative data are usually easy to obtain, they don't tell the full story of where coverage gaps remain. Reasonable assumptions can be made by analyzing uninsured county data (when available) and Medicaid coverage information from the CPS and the ACS. But having access to the administrative data would be more useful for analysis.

FINAL REMARKS

From its inception, CHIP has been very successful at providing access to health insurance for low-income children and families. Uninsured data, both census state-level data and state-produced surveys, have provided valuable information for states to evaluate the program, but they have limits. As CHIP moves forward over the next several years, detailed and more localized data on the uninsured would assist states in locating and addressing the coverage gaps that still exist.

REFERENCES

Blewett, L.A., and Davern, M. (2007). Evaluating federal funding formulas: The State Children's Health Insurance Program. *Journal of Health Policy Politics and Law, 32,* 415-455.

Cook, A., Kenney, G., and Lawton, E. (2010). *Profile of Virginia's Uninsured and Trends in Health Insurance Coverage, 2000-2008.* Washington, DC: Urban Institute.

Georgetown University Center for Children and Families. (2006). Too close to turn back: Covering America's children. Analysis based on data from the National Health Interview Survey.

Holahan, J., Cook, A., and Dubay, L. (2007). *Characteristics of the Uninsured: Who Is Eligible for Public Coverage and Who Needs Help Affording Coverage?* Report from the Kaiser Commission on Medicaid and the Uninsured.

National Academy for State Health Policy. (2008). The CMS August 2007 directive: Implementation issues and implications for State SCHIP programs. *State Health Policy Briefing, 2*(5).

Park, E., and Broaddus, M. (2006). *Fourteen States Face SCHIP Shortfalls This Year Totaling Over $700 Million.* Report from the Center on Budget and Policy Priorities.

12

The Massachusetts Experience: Using Survey Data to Evaluate State Health Care Reform

Sharon K. Long
The Urban Institute

In April 2006, Massachusetts passed landmark legislation that sought to move the state to near-universal health insurance coverage and to improve access to high-quality health care. Following the passage of the legislation, there was substantial concern among key stakeholders in the state about the ability to evaluate the impacts of health care reform with existing state and national data sources. This led the Blue Cross/Blue Shield of Massachusetts Foundation, along with the Commonwealth Fund and the Robert Wood Johnson Foundation, to support a new survey effort dedicated to the evaluation of health care reform—the Massachusetts Health Reform Survey (MHRS). In addition, concerns about the reliability of estimates from an existing state survey of health insurance coverage—the Massachusetts Health Insurance Survey (MHIS)—led to a redesign of that survey to better support the tracking of insurance coverage over time. The MHIS is sponsored by the Massachusetts Division of Health Care Finance and Policy. The Urban Institute, along with Social Science Research Solutions, conducts the MHRS and the MHIS. This paper addresses the lessons learned from the two Massachusetts surveys that may apply to efforts to monitor health insurance coverage of children in the states.

WHY A SEPARATE STATE SURVEY?

Prior to 2006, Massachusetts, like a number of other states, sponsored its own household survey in order to obtain (1) a larger sample for state estimates than is available in national surveys, including larger samples for important subpopulations in the state and for substate geographic areas; (2) information on insurance coverage that addressed the full scope of coverage options available in the state; (3) information on other key issues of relevance to the states (e.g., access to health care and the affordability of health care); and (4) access to the data in a more timely manner to support analyses, policy development, and program design. The MHRS, the new survey effort sponsored by the foundations, had those goals as well, with an additional focus on expanding the information available to track the impacts of the state's 2006 health care reform legislation beyond insurance coverage, such as access to and use of care, health care costs and affordability, the quality of insurance coverage, and support for reform (among other outcomes), and providing timely updates on the impacts of reform.

In 2008, a new source of estimates for uninsurance in Massachusetts became available based on the American Community Survey (ACS), which had added a question on health insurance coverage in the 2008 survey. That survey, which provides a much larger sample size for Massachusetts than is available from other surveys (including the two state-specific surveys), addresses one of the factors that has led states to invest in separate state surveys—sample size. However, the ACS does not address the other needs: state-specific insurance coverage options in the survey questions, information on health care outcomes beyond insurance coverage, and more timely data.[1]

DESIGN OF THE MASSACHUSETTS SURVEYS

Massachusetts Health Insurance Survey

The Massachusetts Division of Health Care Finance and Policy began fielding its health insurance survey in 1998 to provide estimates of the uninsurance rate in the state. The survey was redesigned in 2008 to better position the state to track insurance coverage over time following the passage of the state's 2006 health care reform legislation.[2] Key changes included expanding the survey sample frame to include all residential households (not just those with a land line telephone) and modifying the survey instrument to capture more of the health insurance and health care

[1] For more information on the ACS, see: http://www.census.gov/acs/www/ [July 2010].

[2] See the Massachusetts Division of Health Care Finance and Policy (2007) for information on the early years of the MHRS, and Long et al. (2009) for information on the survey after the 2008 redesign.

options in the state in response to the changes introduced under health care reform.

In order to ensure that the MHIS covered nearly all residents of Massachusetts, a dual sample frame was employed, combining a random digit dial (RDD) landline telephone sample with an address-based sample. The survey, which is available in English, Spanish, and Portuguese, is conducted via telephone, web, and mail, with almost half of the survey respondents completing the survey on the web.

Conducted in the spring of each year, the MHIS collects information on health insurance coverage and basic demographic characteristics for all members of the household. More detailed socioeconomic characteristics and health care information are collected for one randomly selected household member (referred to as the target person in the household) and the other members of the target person's family residing in the household.

The sample size of the MHIS is 4,000 to 5,000 households in each year. The margin of error due to sampling in the 2009 MHIS for estimates based on the full sample of Massachusetts residents is +/–1.5 percentage points at the 95 percent confidence interval.

Massachusetts Health Reform Survey

The MHRS gathers information on nonelderly adults in the state, the primary target population for many of the components of the 2006 health care reform initiative.[3] The MHRS relies on an RDD sample, collecting information on a single, randomly selected adult aged 18 to 64 in each sample household. Like the MHIS, the MHRS is available in English, Spanish, and Portuguese. The key advantages of the MHRS relative to the MHIS are consistent data for the period prior to and following health care reform, a much broader set of questions beyond health insurance coverage, and oversamples of the populations most likely to be affected by the 2006 health care reform legislation—low- and moderate-income adults and uninsured adults.[4] The oversamples of low- and moderate-income adults are obtained through geographic oversamples of households in lower income areas in the state. The oversample of uninsured adults is obtained by screening on insurance status as part of the introduction to the survey.

The MHRS is fielded in the fall of each year. The sample size of the MHRS is 3,000 to 4,000 nonelderly adults in each year, of which 300 to 400 are uninsured. The margin of error due to sampling in the 2009 MHRS for

[3] For more information on the MHRS, see Long (2010).

[4] The 2008 MHRS also included oversamples of racial/ethnic minority populations and populations in some geographic areas in the state.

estimates based on the full sample of nonelderly adults is +/–2.7 percentage points at the 95 percent confidence interval.

COMPARISON OF ESTIMATES OF THE UNINSURANCE RATE FOR MASSACHUSETTS

Uninsurance Prior to Health Care Reform

Estimates of uninsurance in Massachusetts in 2006, just prior to the implementation of key elements of health care reform, are available from both of the state-specific surveys—the MHIS (prior to the redesign) and the MHRS, as well as from national survey efforts, including the Current Population Survey (CPS),[5] the National Health Interview Survey (NHIS),[6] and the Behavioral Risk Factor Surveillance System (BRFSS).[7] The MHIS, the CPS, and the NHIS report coverage for all residents of the state, whereas the BRFSS and the MHRS are limited to nonelderly adults.[8]

The five surveys generated very different estimates of the uninsurance rate in Massachusetts in 2006 (see Figure 12-1). For children, for whom three surveys provided data in Massachusetts in 2006, the uninsurance estimates prior to health care reform ranged from 2.5 percent (MHIS) to 7.1 percent (CPS). For nonelderly adults, for whom five surveys provided data in 2006, the MHIS and the CPS anchored the estimates as well, with the lowest estimate from the MHIS, at 9.2 percent, and the highest estimates from the CPS, at 13.8 percent.

Differences in estimates of the uninsurance rate across surveys are not unique to Massachusetts: the federal government produces multiple estimates of uninsurance based on its own surveys, and estimates from state-sponsored surveys often differ from estimates obtained from the national surveys (Call, Davern, and Blewett, 2007). The differences for Massachusetts from the different surveys reflect many factors, including differences in the sample populations included, differences in the wording of the insurance questions asked, differences in question placement

---

[5] For more information on the health insurance measures in the CPS, see DeNavas-Walt, Proctor, and Smith (2007).

[6] For more information on the NHIS, see Cohen and Martinez (2009).

[7] For more information about the BRFSS, see: http://www.cdc.gov/brfss/about.htm [July 2010].

[8] Estimates of uninsurance rates reported in this section are based on Urban Institute tabulations for the ACS, BRFSS, CPS, and MHRS. Estimates based on the MHIS prior to 2008 are taken from Massachusetts Division of Health Care Finance and Policy (2007), while MHIS estimates for 2008 and 2009 are based on Urban Institute tabulations. Uninsurance estimates from the NHIS are taken from Cohen and Martinez (2006, 2007, and 2009) and Cohen, Martinez, and Free (2008).

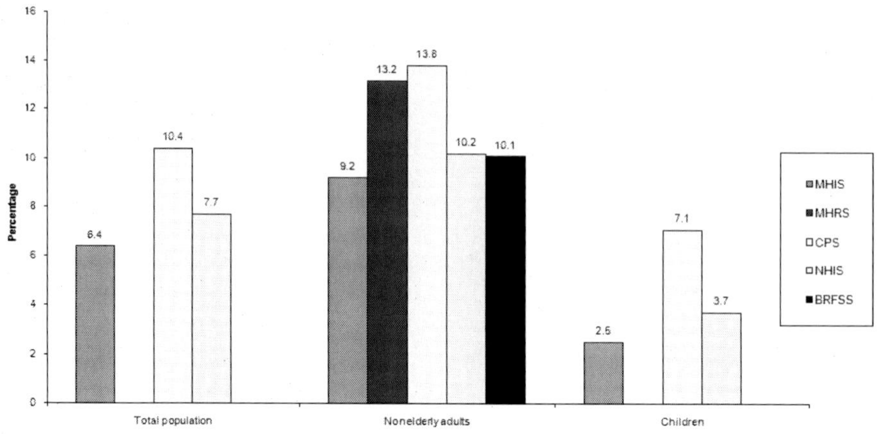

FIGURE 12-1 Estimates of the uninsurance rate in Massachusetts from survey data, 2006.
NOTES: BRFSS = Behavioral Risk Factor Surveillance System, CPS = Current Population Survey, MHIS = Massachusetts Health Insurance Survey, MHRS = Massachusetts Health Reform Survey, NHIS = National Health Interview Survey. Nonelderly adults are defined as aged 19 to 64, except in NHIS and MHRS where they are defined as aged 18 to 64. Children are defined as aged 0 to 18, except in NHIS where they are defined as aged 0 to 17.
SOURCES: Data from Behavioral Risk Factor Surveillance System, the Current Population Survey, the Massachusetts Health Insurance Survey, the Massachusetts Health Reform Survey, and the National Health Interview Survey.

and context, differences in survey design and fielding strategies, and the survey time frame, among other things (Long et al., 2008).[9] In addition, the surveys are based on samples of the Massachusetts population, which, by definition, are subject to error. Consequently, one would not expect different surveys to yield identical estimates of the uninsurance rate in the state.

Uninsurance Under Health Care Reform

Despite the variation in the point estimates of the uninsurance rate in particular years, the general trend in uninsurance in the state is similar

[9] An additional factor that creates some differences in the estimates for children and nonelderly adults is differences in the age cutoffs used to define the two population groups. The MHIS, the CPS, the BRFSS, and the ACS define nonelderly adults as persons aged 19 to 64, and the other surveys (the MHRS and the NHIS) include 18-year-olds in the nonelderly adult category.

across the surveys. This holds true for the total Massachusetts population (see Figure 12-2), nonelderly adults (see Figure 12-3), and children (see Figure 12-4). Note that the NHIS does not report on the uninsurance rate for children in 2007 and 2008 because the relative standard error of that estimate was greater than 50 percent (Cohen and Martinez, 2009).

All of the available surveys show evidence of a substantial drop in uninsurance in Massachusetts since health care reform began in 2006. For the overall population, both the NHIS and the CPS show a drop of about 5 percentage points in the uninsurance rate between 2006 and 2008. For nonelderly adults, the estimate of the drop in the uninsurance rate ranges from 5 percentage points (BRFSS) to 9 percentage points (MHRS), with both the NHIS and the CPS showing a drop of about 6 percentage points. These findings are in contrast to trends in the nation as a whole, in which there was little change between 2006 and 2008 in the high levels of uninsurance (data not shown). Based on tabulations from the NHIS and the CPS, uninsurance for the total U.S. population remained steady at

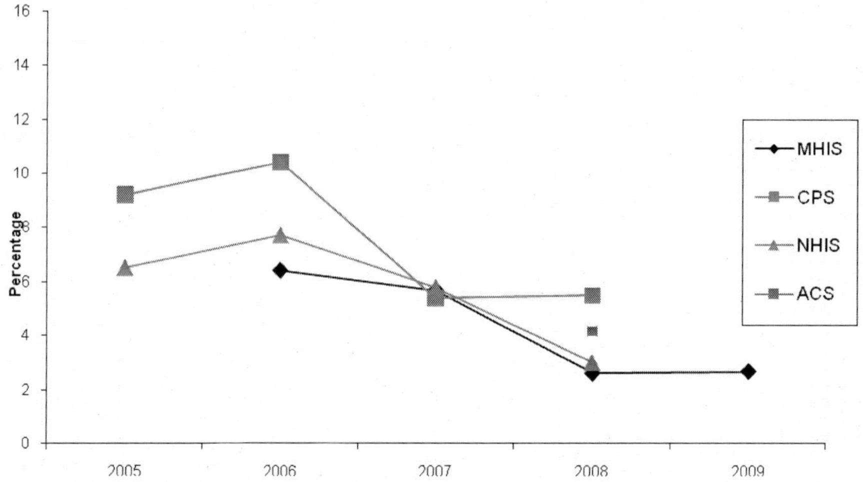

FIGURE 12-2 Trends in uninsurance for the total population in Massachusetts from survey data, 2005-2009.
NOTES: ACS = American Community Survey, CPS = Current Population Survey, MHIS = Massachusetts Health Insurance Survey, NHIS = National Health Interview Survey. Nonelderly adults are defined as aged 19 to 64, except in NHIS and MHRS where they are defined as aged 18 to 64. Children are defined as aged 0 to 18, except in NHIS where they are defined as aged 0 to 17.
SOURCES: Data from the American Community Survey, the Current Population Survey, the Massachusetts Health Insurance Survey, and the National Health Interview Survey.

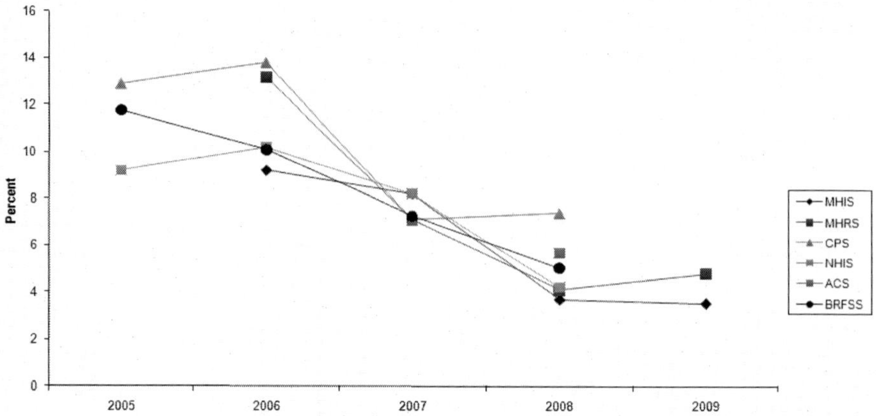

FIGURE 12-3 Trends in uninsurance for nonelderly adults in Massachusetts from survey data, 2005-2009.
NOTES: ACS = American Community Survey, BRFSS = Behavioral Risk Factor Surveillance System, CPS = Current Population Survey, MHIS = Massachusetts Health Insurance Survey, MHRS = Massachusetts Health Reform Survey, NHIS = National Health Interview Survey. Nonelderly adults are defined as aged 19 to 64, except in NHIS and MHRS where they are defined as aged 18 to 64.
SOURCES: Data from the American Community Survey, Behavioral Risk Factor Surveillance System, the Current Population Survey, the Massachusetts Health Insurance Survey, the Massachusetts Health Reform Survey, and the National Health Interview Survey.

about 15 percent between 2006 and 2008, with uninsurance at 20 percent for nonelderly adults in both surveys. The uninsurance rate for children remained steady at 9 percent based on the NHIS, however, it dropped from 12 to 10 percent in the CPS over the same period.

Looking in more detail at the estimates of the uninsurance rate for children in 2008, one sees that all three of the available surveys report very low levels of uninsurance for children in Massachusetts (see Figure 12-5). For children overall and for children in each of the income groups, the highest estimate of the uninsurance rate is reported by the CPS and the lowest estimate by the MHIS, with the ACS between the two. For example, the CPS estimates that 3.4 percent of Massachusetts children were uninsured in 2008, compared with an estimate of 2.1 percent in the ACS and 1.2 percent in the MHIS.

Over time, estimates of the uninsurance rate based on the CPS have tended to be higher than those of other surveys in Massachusetts. This pattern is consistent with studies for other states, which generally have found uninsurance estimates higher in the CPS than in state-specific sur-

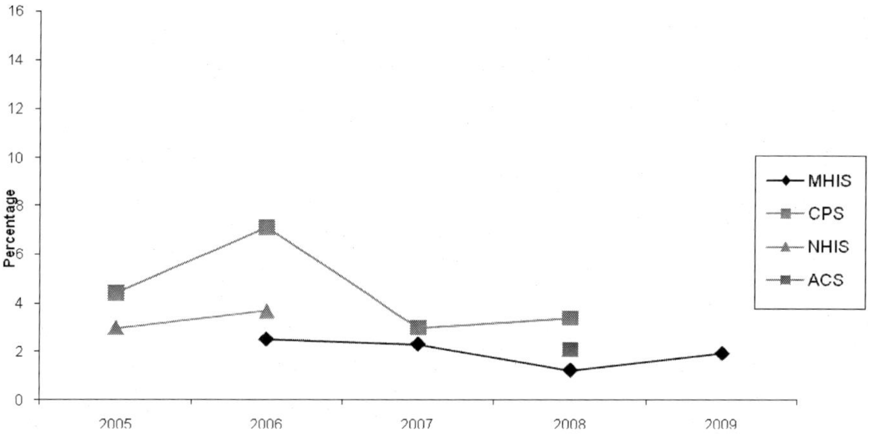

FIGURE 12-4 Trends in uninsurance rates for children in Massachusetts from survey data, 2006-2009.
NOTES: ACS = American Community Survey, CPS = Current Population Survey, MHIS = Massachusetts Health Insurance Survey, NHIS = National Health Interview Survey. Children are defined as aged 0 to 18, except in NHIS where they are defined as aged 0 to 17.
SOURCES: Data from the American Community Survey, the Current Population Survey, the Massachusetts Health Insurance Survey, and the National Health Interview Survey.

veys (State Health Access Data Assistance Center, 2007). A key difference between the CPS and the other surveys reported here is the insurance question used. The CPS, which is fielded in March of each year, asks about insurance coverage over the prior calendar year, whereas the other surveys ask about insurance coverage at the time of the survey. However, the available evidence suggests that respondents are not reporting coverage for the prior calendar year accurately in the CPS (Klerman et al., 2009), raising questions about the appropriate interpretation of the insurance measure in the CPS.

LESSONS LEARNED

The two Massachusetts surveys have provided valuable information as the state has implemented its landmark health care reform initiative, including timely feedback on the impacts of reform on insurance coverage, health care access and use, health care affordability, support for health care reform and other measures, and information on key population subgroups in the state. The range of analyses supported by those

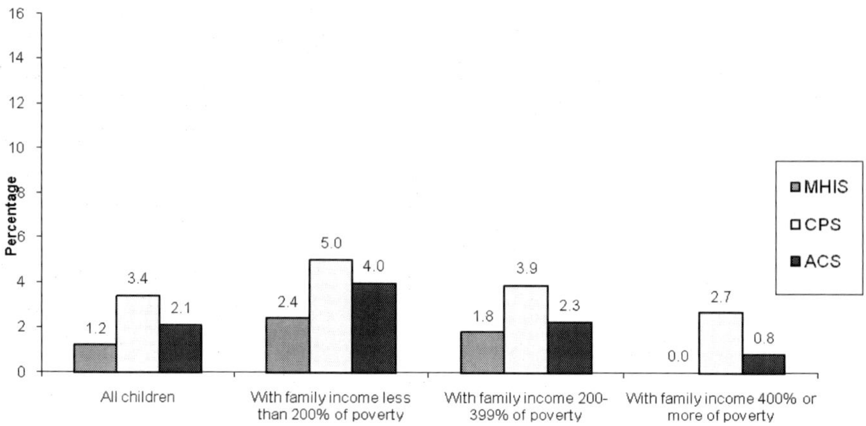

FIGURE 12-5 Estimates of the uninsurance rate for children in Massachusetts from survey data, by family income, 2008.
NOTES: ACS = American Community Survey, CPS = Current Population Survey, MHIS = Massachusetts Health Insurance Survey. Children are defined as aged 0 to 18.
SOURCES: Data from the American Community Survey, the Current Population Survey, and the Massachusetts Health Insurance Survey.

surveys could not have been accomplished with any other available data sources. However, the surveys have faced a number of challenges:

- ***Maintaining funding over time.*** The MHRS began with support from three foundations; however, with the economic downturn and changes in foundation priorities, two of the foundations decided not to continue supporting the survey effort in 2009. The MHIS, which is sponsored by the state, is facing cutbacks in 2010 as Massachusetts, like every other state, is facing serious financial difficulties.
- ***Limited ability to support research on survey methods.*** Given the relatively limited funds available for the surveys, few resources have been available to support research on survey methods. Both surveys have undertaken new strategies to try to address specific survey goals (e.g., including cell phone–only households in the MHRS, increasing survey response rates), which researchers have not been able to evaluate. Of particular relevance, the MHIS introduced a new, dual frame survey design that relies heavily on survey respondents completing the survey on the web. It would be very useful to conduct a controlled experiment to see if responses

obtained via the web differ from those obtained using a telephone interviewer. This is an area in which federal funding to improve survey research methods would be valuable.

- ***Limited ability to share the data.*** Limits on the available survey funds have also prevented the creation, dissemination, and support of public-use data files based on the two surveys. This, in turn, limits the use of the survey data for additional research projects. This is another area in which federal funding would be valuable to help support the creation, maintenance, and dissemination of public-use data files.
- ***Comparisons to other states.*** Although both surveys provide valuable data on Massachusetts, the lack of comparable national data makes it difficult to compare Massachusetts with the rest of the country on a number of important outcomes. Such information would be valuable to place the findings for Massachusetts in the context of other states, as well as to support stronger evaluation designs than are possible with data for a single state.

Although there is likely to be a continuing role for state-sponsored surveys to address issues of particular policy relevance in each state, a number of strategies would increase the value of existing national surveys for state-specific studies. These include

- providing much larger state and local samples, overall and for key population groups (including children);
- making state identifiers available outside research data center settings;
- providing more geocoding of local areas;
- adding state-specific program names to health insurance coverage questions;
- expanding survey content to include questions on access, use, and costs of care, along with other issues of relevance to national health care reform; and
- making data files available more quickly and in user-friendly formats to facilitate their use by state analysts.

ACKNOWLEDGMENT

This paper reflects the views of the author and does not necessarily represent the views of the Urban Institute, its sponsors, or its trustees.

REFERENCES

Call, K.T., Davern, M., and Blewett, L.A. (2007). Estimates of health insurance coverage: Comparing state surveys with the Current Population Survey. *Health Affairs, 26*(1), 269-278.

Cohen, R.A., and Martinez, M.E. (2006, June). *Health Insurance Coverage: Early Release of Estimates from the National Health Interview Survey, 2005.* Division of Health Interview Statistics, National Center for Health Statistics. Available: http://www.cdc.gov/nchs/data/nhis/earlyrelease/insur200606.pdf [May 2010].

Cohen, R.A., and Martinez, M.E. (2007, June). *Health Insurance Coverage: Early Release of Estimates from the National Health Interview Survey, 2006.* Division of Health Interview Statistics, National Center for Health Statistics. Available: http://www.cdc.gov/nchs/data/nhis/earlyrelease/insur200706.pdf [May 2010].

Cohen, R.A., and Martinez, M.E. (2009, June). *Health Insurance Coverage: Early Release of Estimates from the National Health Interview Survey, 2008.* Division of Health Interview Statistics, National Center for Health Statistics. Available: http://www.cdc.gov/nchs/data/nhis/earlyrelease/insur200906.pdf [May 2010].

Cohen, R.A., Martinez, M.E., and Free, H.L. (2008, June). *Health Insurance Coverage: Early Release of Estimates from the National Health Interview Survey, 2007.* Division of Health Interview Statistics, National Center for Health Statistics. Available: http://www.cdc.gov/nchs/data/nhis/earlyrelease/insur200806.pdf [May 2010].

DeNavas-Walt, C., Proctor, B.D., and Smith, J. (2007). *Income, Poverty, and Health Insurance Coverage in the United States: 2006.* U.S. Census Bureau, August. Available: http://www.census.gov/prod/2007pubs/p60-233.pdf [May 2010].

Klerman, J.A., Davern, M., Call, K.T., Lynch, V., and Ringel, J.D. (2009). Understanding the Current Population Survey's insurance estimates and the Medicaid "undercount". *Health Affairs, 28*(6), w991-w1001.

Long, S.K. (2010). *The Massachusetts Health Reform Survey.* Washington, DC: The Urban Institute. Available: http://www.urban.org/UploadedPDF/411649_mass_reform_survey.pdf [May 2010].

Long, S.K., Zuckerman, S., Triplett, T., Cook, A., Nordahl, K., Siegrist, T., and Wacks, C. (2008). *Estimates of the Uninsurance Rate in Massachusetts from Survey Data: Why Are They So Different?* Boston: Massachusetts Division of Health Care Finance and Policy, August. Available: http://www.mass.gov/Eeohhs2/docs/dhcfp/r/pubs/08/est_of_uninsur_rate.pdf [May 2010].

Long, S.K., Triplett, T., Dutwin, D., and Sherr, S. (2009). *2008 Massachusetts Health Insurance Survey, Methodology Report.* Boston: Executive Office of Health and Human Services. Available: http://www.mass.gov/Eeohhs2/docs/dhcfp/r/survey/08his_methodology_rev.pdf [May 2010].

Massachusetts Division of Health Care Finance and Policy. (2007). *Massachusetts Household Survey on Health, Insurance Status, 2007.* Boston: Executive Office of Health and Human Services. Available: http://www.mass.gov/Eeohhs2/docs/dhcfp/r/pubs/08/hh_survey_07.ppt [May 2010].

State Health Access Data Assistance Center (SHADAC). (2007). *State Health Insurance Coverage Estimates: A Fresh Look at Why State Survey Estimates Differ from the CPS.* Issue Brief #12, November. Minneapolis: University of Minnesota. Available: http://www.shadac.org/files/IssueBrief12.pdf [May 2010].

13

Small-Domain Estimation of Health Insurance Coverage

Brett O'Hara and Mark Bauder
U.S. Census Bureau

The Census Bureau's Small Area Health Insurance Estimates (SAHIE) Program uses small-area techniques to estimate health insurance coverage. These model-based estimates generally have lower variances than survey-only (i.e., nonmodel-based) estimates because the modeling incorporates covariates that are related to health insurance coverage. The SAHIE program is an outgrowth of its sister program, Small Area Income and Poverty Estimates (SAIPE). The primary purpose of SAIPE is to make model-based estimates of the number of poor, school-age children in a school district. For over a decade, the U.S. Department of Education has used these estimates as the major component of its allocation formula for dispersing funds from Title I of the Elementary and Secondary Education Act. From SAIPE, there was a knowledge base that the SAHIE program has benefited from and used.

In 2005, the SAHIE program published health insurance coverage estimates for the first time. This first release of data was to determine the feasibility of producing substate estimates when a limited set of estimates was published for only two age cohorts. The survey data used came from the Annual Social and Economic Supplement of the Current Population Survey (CPS ASEC). At the time, the CPS ASEC was the only survey that could produce health insurance estimates for states. State estimates were designed to be reliable when 3 years of CPS ASEC were combined. However, the survey could not produce reliable estimates for all counties. Not all counties are in the survey, and most counties in the survey have small sample sizes. The SAHIE team established that it was feasible to produce estimates for all counties by using a methodology similar to SAIPE.

In 2007, the SAHIE program published state estimates of the prevalence of uninsured low-income women by age, race, and Hispanic origin. Low income is defined in terms of family income divided by the poverty threshold. The subpopulations of interest were determined by a sponsor of the SAHIE program, the Centers for Disease Control and Prevention's (CDC's) National Breast and Cervical Cancer Early Detection Program (NBCCEDP). Its funding of the SAHIE program made it possible to produce estimates of income-eligible women by age group (18-64, 40-64, and 50-64). Income eligibility for this benefit varies by state. Most state programs chose income eligibility at 0-200 or 0-250 percent of the poverty threshold. From these estimates, the SAHIE team evaluated whether the state model could meet the CDC requirements. State and county estimates of health insurance coverage were published in 2008 that met the requirements. The SAHIE model was later modified so that it had the functionality to support many more income categories and more age categories. However, the 3-year average of the CPS ASEC health insurance coverage data could not support model-based estimates of more income categories or additional age groups (Small Area Health Insurance Estimates Team, 2008).

Currently, the SAHIE program publishes state health insurance coverage estimates by age, sex, race, Hispanic origin (i.e., demographic characteristics), and by income categories (both 0-200 percent and 0-250 percent of the poverty threshold and the total poverty universe). For counties, SAHIE produces estimates by age, sex, and income categories (either 0-200 percent or 0-250 percent of the poverty threshold and the total poverty universe). Estimates from this model were released in 2009 with no significant changes from the year before. Before 2009, all of the SAHIE estimates were labeled "experimental." For 2009, the SAHIE program had evolved to use a production model and production health insurance coverage estimates.

One of the SAHIE program goals is to accelerate its production process to publish estimates in a timelier manner. Currently, the modeled estimates are released in the summer, based on the most current 3 years of the CPS ASEC. The program goal is to release estimates based on current data in the early winter.

Evaluation of various model forms to produce SAHIE estimates is ongoing. The model-based estimates had smaller confidence intervals than the survey-only estimates for uninsured low-income children under age 18. In addition, the two estimates for a given state were close (O'Hara, 2008). The SAHIE numbers can also be used as a low-variance denominator for participation (or nonparticipation) rates for the uninsured populations. This empirical application estimated participation rates using administrative data from NBCCEDP (the numerator) and uninsured low-income women ages 40-64 at the state and county levels (the denominator) (O'Hara, Tangka, and Bauder, in review).

The SAHIE program is currently exploring how to take best advantage of the American Community Survey (ACS) data. Full production of the ACS in 2006 allowed it to replace the decennial census sample data (i.e., the long form). Beginning in 2008, the ACS included a health insurance question. As a result, there are annual ACS health insurance estimates for all geographic areas with a population of 65,000 or more (U.S. Census Bureau, 2009). The ACS contains approximately 30 times the number of addresses as a single-year CPS ASEC. Direct survey estimates, based on a 5-year accumulation from the ACS, for geographic areas with population of 20,000 or less will not be available until 2013.

This paper assesses an ACS-based SAHIE model in two ways. First, we compare the variances of state estimates of a CPS-based SAHIE model and an ACS-based SAHIE model. Second, the model is modified to obtain ACS model-based estimates for more income categories. The gains from model-based state estimates increase as the number of domains increases (e.g., the income categories). As yet, county-level estimates have not been explored. These potential changes to the model, based on the strengths of the ACS, are intended to create more useful or refined estimates of the uninsured populations for policy makers and other stakeholders.

BACKGROUND

Uninsured children are likely to use emergency room visits as a form of "free" primary care and to put off any use of medical services until a need must be met (Grumbach, Keane, and Bindman, 1993). Even when an uninsured child uses other care, access to and utilization of the care are lower than that of children with health insurance coverage (Newacheck, Hughes, and Stoddard, 1996). Children without health insurance coverage are more likely to have unmet medical needs that affect long-term health status compared with insured children (Newacheck et al., 2000). Being uninsured increases the risk of having high out-of-pocket expenses for medical services. As a result, families with uninsured children are more likely to become impoverished when the child has a health event (O'Hara, 2004).

The first nationwide means-tested health insurance program was the Medicaid program. Created in 1965, Medicaid provides full and comprehensive health insurance coverage (as well as partial benefits). The target populations are poor and near-poor children, the disabled, and the elderly (Centers for Medicare & Medicaid Services, 2010). In 1997, a new means-tested health insurance program was enacted—the Children's Health Insurance Program (CHIP). It targets uninsured children who do not qualify for Medicaid but whose family income is too low to pay insurance premiums (Dubay et al., 2007). Of the children whose family income fell between 133 and less than or equal to 200 percent of poverty, 22.4 percent were uninsured in 1996 and 14.6 percent were uninsured in 2008 (authors' own calculation using the CPS).

The federal CHIP allocation to the states was dependent on several factors, one of which was an estimate of the state's number of uninsured children under age 19 who are at or below 200 percent of poverty. Until recently, the only survey that produced those estimates was the CPS; however, it was ill suited for making state allocations for CHIP because the survey had a sample size too small to produce accurate estimates of the target population (Czajka and Jabine, 2002; National Research Council, 2002). The Medicare, Medicaid, and S CHIP Balanced Budget Refinement Act of 1999 provided new funding to increase the sample size of the CPS in order to improve the reliability of the estimates (P.L. 106-113, 1999). The expansion of the CPS was implemented in collection year 2001 (U.S. Census Bureau, 2002).

Even with the CPS expansion, hereafter referred to as the CPS ASEC, much of the states' year-to-year fluctuations in the estimates were still large, probably due to sampling error and not real change. Only 10 states have a sample size that results in estimates of more than 100 uninsured children at or below 200 percent of poverty (Blewett and Davern, 2007). Because of this small sample size, one suggestion has been to use model-based estimates of the target population that combine the ACS and CPS information (Blewett and Davern, 2007). In the Children's Health Insurance Program Reauthorization Act of 2009 (CHIPRA), the allocation formulas are based on population estimates, not uninsured or low-income estimates (P.L. 111-3, 2009). It is possible that this decision was acknowledgment of the sampling error in the CPS ASEC. Under CHIPRA, the Census Bureau is required to rank the states in terms of their rates of uninsured low-income children and continue to produce population estimates of children under age 19. This part of the law is to determine "high-performing States," that is, states with high insured rates. The Census Bureau was also required to determine whether the ACS or the CPS ASEC gives the best population estimates.

In September 2009, the ACS released its first set of health insurance estimates. Sample size is not a constraint for producing ACS estimates for states and many substate geographic areas. In the Patient Protection and Affordable Care Act of 2010 (PPACA), there are provisions that the Census Bureau will provide state-level estimates of the uninsured using the ACS and provide "data at the smallest geographic level" (P.L. 111-148, 2010).

PPACA made a correction to CHIPRA concerning population estimates; the law was changed to strike out "population estimates" and replace the phrase with "a high-performing State." The Health Care and Education Reconciliation Act of 2010 (HCERA) requires the Census Bureau to rank the states in terms of their rates of uninsured people (P.L. 111-152, 2010).

Although it is not a requirement for the Census Bureau, the PPACA has new Medicaid requirements (all people at or below 133 percent of poverty are income-eligible), and establishes state-sponsored health insurance exchanges (all people between 133 percent and less than or equal to 400 percent of poverty are eligible for premium assistance). The level of premium assistance varies by the income-to-poverty ratio. For example, if the family income is between 150 and less than or equal to 200 percent of poverty, the family's share of the premium would be less than or equal to 6.3 percent of the family's income (Kaiser, 2010).

The SAHIE program has been responsive to stakeholders' needs. It has always emphasized uninsured populations for the allocation of funds, outreach, and programmatic evaluations. For example, the NBCCEDP uses the SAHIE estimates to direct outreach funds to underserved counties. SAHIE state and county estimates of uninsured children can be used in the same way, as well as providing "data at the smallest geographic level." The focus of the SAHIE program has changed somewhat given CHIPRA, the PPACA, and the HCERA. To be responsive to stakeholders' needs, the SAHIE program is researching the feasibility of providing policy-relevant income-to-poverty ratios. Using the ACS, this research is possible.

DATA

The SAHIE program constructs statistical models that relate surveys, population estimates, and administrative records to produce health insurance estimates. The CPS ASEC provides annual national and state estimates based on a sample of about 100,000 addresses. All health insurance coverage and income questions refer to the previous year. Of the more than 3,100 counties in the United States, about 1,200 counties are in the sample. For the CPS-based SAHIE model, a 3-year average is used (Small Area Health Insurance Estimates, 2009). The CPS ASEC used in this evaluation are the collection years of 2007, 2008, and 2009. The ACS provides annual national, state, and substate estimates based on a sample of about 3 million addresses. Health insurance questions refer to coverage that day. Income questions refer to income over the last 12 months, and annual estimates cover a period of 23 months. A single year of survey data is used for the ACS-based SAHIE model; the 2008 ACS was used for this evaluation. For both surveys, family size is determined at the time of the interview.

Dividing family income by the appropriate poverty threshold determines a family's income group. For the ACS, both family income and the poverty threshold are adjusted monthly for inflation (U.S. Census Bureau, 2009). The survey data are tabulated by the number of uninsured by sex, age, race, and Hispanic origin categories (non-Hispanic white, non-Hispanic black, non-Hispanic people of other races or multiracial, Hispanic), and relevant income-to-poverty ratios. Both the CPS ASEC and the ACS are tabulated from microdata and are not published state estimates.

The covariates of income-to-poverty ratios are different from those for health insurance coverage. Covariates for income-to-poverty ratios include the estimates from the 2000 census sample data tabulated for the same categories as the survey data as well as other characteristics reported in the census sample data. The number of federal tax exemptions is calculated by age groups (children, nonelderly adults, and elderly adults) in various income-to-poverty ratios. The number of Supplemental Nutrition Assistance Program (SNAP) participants is also a strong predictor of income groups. The number of Medicaid and CHIP participants (for categories defined by age and sex) and the number of people employed in different-sized firms as shown in administrative records (the County Business Patterns) are strong predictors of health insurance. These predictors are aggregated to the county and state levels (O'Hara, 2008).

METHODS

The SAHIE program has developed a statistical model for producing estimates of numbers and proportions of insured and uninsured within income groups. For the geographic areas and domains of interest, survey-only estimates of the numbers in income groups, as well as numbers of insured, are often not adequate for users' needs. Thus, the SAHIE model produces model-based estimates of both the numbers in income groups within geographic and demographic groups and the proportions and numbers insured within those income groups. Survey-only estimates of numbers in the income groups and the numbers insured and uninsured are modeled in terms of the quantities to be estimated. The model treats data from some of the administrative data in a way similar to survey estimates. They are modeled as random with distributions depending on the quantities to be estimated. Other data, in particular data from the County Business Patterns, are in the model as fixed predictors.

The model is a combination of two hierarchical models, corresponding to the proportions in income groups and proportions insured. The model is estimated as a fully Bayesian model. Vague prior distributions are assigned to unknown parameters. Point estimates are posterior means, and posterior variances are used to construct confidence intervals. The model is estimated using Markov chain Monte Carlo methods.

Model Details

The current SAHIE model is similar to common small-area models. A basic and well-known example is the Fay-Herriot model (Fay and Herriot, 1979). Suppose that for area, i, Y_i is the variable being estimated, and $\hat{Y}_i$ is its survey estimate. In the Fay-Herriot model, given some predictors and parameters, Y follows a normal linear model and, conditional on Y_i and parameters, and $\hat{Y}_i$ is unbiased and normally distributed. The Fay-Herriot model can be written:

$$\begin{aligned} \hat{Y}_i \big| Y_i, \theta &\sim N(Y_i, v_i^S(\theta)) \\ Y_i \big| x_i, \beta, \theta &\sim N(x_i^T \beta, v_i^M(\theta)). \end{aligned}$$

The variances v_i^M and v_i^S are *model* and *sampling* variances. θ contains any parameters on which these variances depend. Common simplifying assumptions are that the Y_i's are independent given the x_i's and parameters, the $\hat{Y}_i$'s are independent given the Y_i's and parameters, and that the model variances are the same for all i. It is also common to plug in survey or model-based estimates of the sampling variances and treat them as known.

The SAHIE model shares the hierarchical form of the Fay-Herriot model. But there are several differences between the current SAHIE model and the basic Fay-Herriot model described above:

- There are two parts to the SAHIE model, each of which has a structure similar to a Fay-Herriot model. One part is for estimating proportions in income categories, the other is for estimating proportions insured.
- Transformations of the proportions in the income categories and the proportions insured follow linear models, given predictors and parameters, whereas direct estimates of the *untransformed* proportions or numbers are unbiased and normal.
- The two parts of the SAHIE model are "multivariate." In addition to the survey estimates, other data are treated as random and are modeled by their distributions conditional on the true numbers and proportions.
- Sampling variances are modeled, rather than treated as known.

Basic SAHIE model

We describe a basic version of the SAHIE model below, using the following notation:

- a indexes state/age/race/sex or county/age/sex, and i indexes income group.
- p^I_{ai} is the proportion of those in domain a who are in income group i.
- p^H_{ai} is the proportion with health insurance within domain a, income group i.
- N^I_{ai} and N^H_{ai} are the corresponding number in the income group and number insured.
- POP_a is the demographic population estimate for domain a, assumed known.

The proportions and numbers are related by

$$N^I_{ai} = p^I_{ai}\ POP_a \text{ and } N^H_{ai} = p^H_{ai}\ N^I_{ai}.$$

A basic version of the SAHIE model is as follows:

- Survey estimates of the proportions in the income group and of the proportions with health insurance, given the true proportions, are normal, unbiased, and have variances that depend on parameters to be estimated. Let p denote either a proportion in an income group or a proportion insured, with corresponding survey estimate $\hat{p}$.

$$\hat{p}_{ai} \sim N(p_{ai}, v^S_{ai}) \quad \text{with } v^S_{ai} = \lambda_1 \frac{p_{ai}(1-p_{ai})}{S^{\lambda_2}_{ai}}.$$

- Census 2000 estimates, tax exemptions, SNAP participation, and combined Medicaid and CHIP participation are random. Each is normally distributed, given the true numbers in income groups and numbers insured. Except for 2000 census data, these data are not at the level of the survey estimates or the p_{ai}'s. Instead, they correspond to aggregates over base-level domains. Each has an associated parametric function for its mean and variance. Let Z be data from one of these four sources, and let t index its records. In a typical case, we have

$$Z_t \sim N\ (\tilde{Z}_t, \text{var}(Z_t))\ \tilde{Z}_t = \alpha N^*_t \text{ and } \text{var}(Z_t) = \eta 1 \cdot (\tilde{Z}_t)^{\eta 2}.$$

Here, N^*_t is a sum of numbers in income groups or numbers insured over a set of indices a, i that correspond to record t.

- Transformations of the proportions in the income groups and of the proportions insured follow normal linear models. For each of the two proportions estimated, there is a matrix of predictors, x, a vector of regression coefficients, β, a model variance ν^m, and a link function L.

$$p_{ai} = L^{-1}(\mu_a) \quad \mu_{ai} = x_{ai}^T\beta + \varepsilon_{ai} \quad \text{with } \varepsilon_{ai} \sim N(0, \nu^m)\ iid$$

Production and test models differ in details from the basic model described above. Parameters (the λ's, α's, and η's above) can vary by demographic group. Models may have nonlinear functions for means and may assume distributions other than normal. There can be random effects in the regressions and mean functions. For specifics about production SAHIE models, see Bauder et al. (2008).

Model Evaluation

There is no "gold standard" relative to which the accuracy of SAHIE model estimates can be checked. There is no population for which the true numbers insured and uninsured are known, to which we could apply and test the model. At best, we can test the model assumptions by how well the model fits the observed data.

We use several techniques to assess the fit of the model. In addition to checking the plausibility of parameter estimates, we rely strongly on residual analysis and posterior predictive checks. In the process of estimating the model, we generate predictions of means and variances of the survey and administrative data. We can then check whether the observed data are consistent with these predicted means and variances. We also check for patterns in the residuals. Posterior predictive checks involve generating replicate data from the estimated distributions of observed data. We then check whether these replicate data resemble the observed data.

Model Weaknesses

There are several weaknesses in the modeling strategy that we hope to address in future research.

- **Assumptions of independence of observations.** It is common to assume independence among observations at the level of the data that are modeled. There are practical reasons for making these assumptions, since estimation can become much more complex and difficult without them. It may be possible to relax these assumptions. At the least, the effect of such assumptions should be investigated.

- **Assumptions about unobserved variables.** The income and insurance parts of the SAHIE model are hierarchical, with the variables of interest occupying a different level than observed data in the hierarchy. Because these variables of interest are unobserved and inferable only by their "effect," assumptions about their distributions are harder to justify and harder to verify than, say, assumptions about survey data. For example, we assume that a transformation of the proportion insured, given some predictors and parameters, follows a regression model with errors that are independent and have the same variance. This kind of assumption is common in small-area models but is worth investigating.

RESULTS

The SAHIE program has produced valuable estimates, using the CPS ASEC as the survey data. In this section, we report results from analyses of state-level estimates using the ACS 2008 data. We investigate whether we expect SAHIE estimates to be better using 1 year of the ACS data rather than 3 years of the CPS ASEC data and whether modeling for smaller domains is feasible using the ACS data.

Comparison of Models

Here, we compare the reliability of estimates from a CPS-based and an ACS-based model. We use models that are as similar as possible, except for the survey data. We use estimates from a preproduction estimation of the model using the CPS ASEC for collection years 2007, 2008, 2009 and from the corresponding ACS model for 2008.

We concentrate on one group of interest, uninsured children under age 19 at or below 200 percent of poverty by health insurance coverage. Variances or standard deviations of estimates are measures of uncertainty, so they are a reasonable basis for comparing estimates from different models. We calculated the ratios of posterior standard deviations from an ACS-based model over those from a CPS-based model. In Figure 13-1, we plot these ratios as the ACS-based standard deviation over the corresponding CPS-based one. The ratio of standard deviations is approximately the ratio of corresponding confidence intervals. The ACS-based estimates are generally better. At the median ratio of 0.78, a confidence interval from the ACS-based model will be approximately 20 percent shorter than an equivalent one from the CPS-based model. In only three

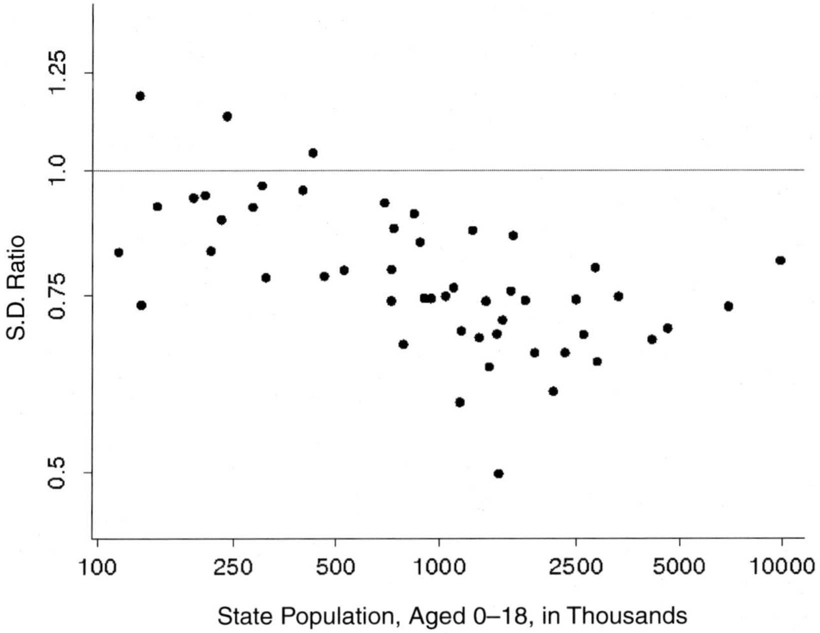

FIGURE 13-1 Ratios of posterior standard deviations from ACS-based and CPS-based models.
NOTES: For the number of uninsured children under 19 with family income less than or equal to 200 percent of poverty. The standard deviation from the ACS-based model is the numerator. The axes are on log scale.
SOURCE: Estimates are from the 2008 ACS-based and the 2007 CPS-based SAHIE models.

states, Idaho, Rhode Island, and Wyoming, do the ACS-based estimates have larger standard deviations.

Estimating More Income Groups

An anticipated advantage of using the ACS rather than the CPS ASEC data is that we will be able to produce reliable model-based estimates for smaller domains. We report here results from some early analyses to assess this possibility.

We modeled data and estimated the numbers and proportions insured and uninsured for the full cross-classification of six income groups: 0-133, 133-150, 150-200, 200-250, 250-400, and over 400 percent of poverty; four age groups: 0-18, 19-39, 40-49, and 50-64; four race/ethnicity groups: non-Hispanic white, non-Hispanic black, non-Hispanic people of other races or multiracial, and Hispanic; and both sexes.

To illustrate the results, we consider a small income group—family income between 150 and less than or equal to 200 percent of poverty. We again consider children under age 19. The model-based estimates are aggregations over the race/ethnicity groups and the two sexes. We assess the value of model-based estimates by looking at the improvement in standard deviations compared with those of survey-only estimates.

We first confirm that the model estimates are consistent with the ACS direct estimates at this moderate level of aggregation. The estimates should be similar if each estimate is measuring the truth and the populations being estimated are large enough that the survey estimate is fairly reliable. A proper significance test is not possible here because the modeled estimates are based on the ACS direct estimates, so they are correlated, and we do not know the correlation. We can informally check that the estimates are close. Figure 13-2 shows the ACS estimates plotted against the model-based estimate for states for the number of uninsured children under age 19 with family income between 150 and less than or equal to 200 percent of poverty. The estimates are close. We also note that

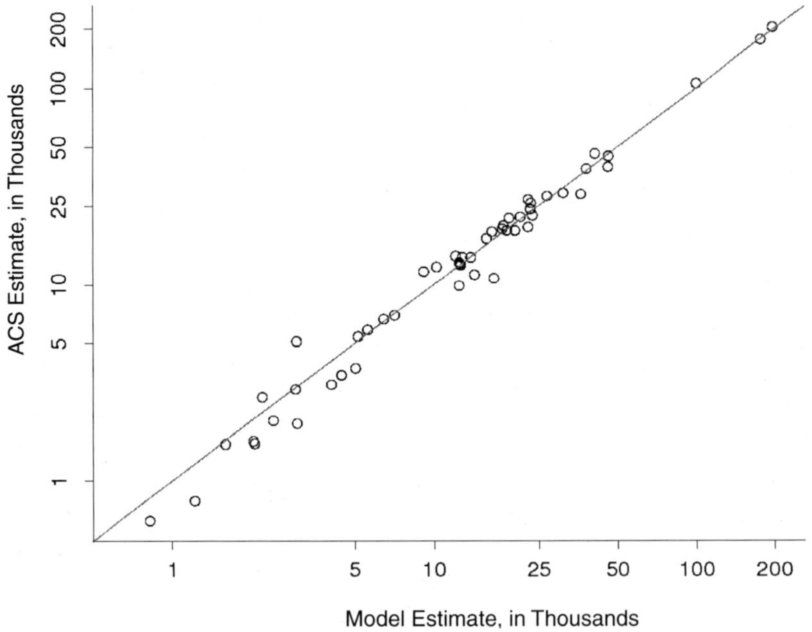

FIGURE 13-2 Plot of the ACS direct estimates versus the ACS model-based estimates, for states, of the number of uninsured children under age 19 with family income between 150 and less than or equal to 200 percent of poverty.
SOURCES: Estimates are from the 2008 ACS survey and the 2008 ACS-based SAHIE model.

the largest percentage differences are for South Dakota (71 percent), Louisiana (36 percent), and New Hampshire (36 percent). For only three states is the difference larger in magnitude than twice the ACS standard error.

Figure 13-3 shows the ratios of model estimate standard deviations to estimates of ACS-only standard deviations for state estimates of uninsured children under age 19 in the income group of 150-200 percent of poverty. All ratios are less than 1, so the model-based estimates are always "better." As expected, there is less advantage to the model-based estimates as the population (and presumably the sample size) increases, because the survey estimates become more reliable. The maximum ratio, representing the smallest improvement, is 0.82 for California. The minimum ratio, representing the largest improvement, is 0.20 for South Dakota. Thus, for this group of estimates, confidence intervals from the model-based esti-

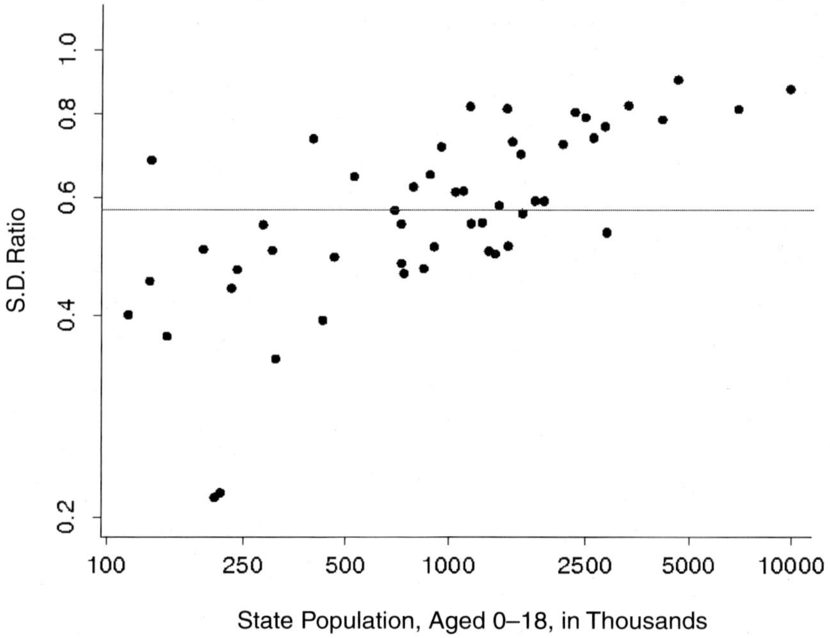

FIGURE 13-3 Ratios of standard deviations of model-based estimates to estimates of standard deviations of survey-only estimates.
NOTES: Standard deviations of the state estimates of the number uninsured children under age 19 with family income between 150 and less than or equal to 200 percent of poverty. The horizontal line is at the median of 0.54. Axes are on log scale.
SOURCES: Estimates are from the 2008 ACS survey and the 2008 ACS-based SAHIE model.

mates are approximately one-fifth to four-fifths the length of those from the survey-only estimates.

A common measure of the goodness of an estimate is the coefficient of variation (CV), defined as the estimate's standard deviation divided by its mean. We estimate the CV of an ACS estimate by dividing the estimate of its standard deviation by the estimate itself. Figure 13-4 shows the estimated CVs for the model estimates and the ACS estimates, for state estimates of the number of uninsured children under age 19 with family income between 150 and less than or equal to 200 percent of poverty. The states are ordered by increasing population under age 19. When CVs are high, estimates are likely to fluctuate relatively more over time because of sampling error than when they are low.

Table 13-1 shows summary statistics for CVs for these SAHIE model estimates of the number uninsured at the level of the data for the domains at the detailed modeling. The table also contains the same statistics for

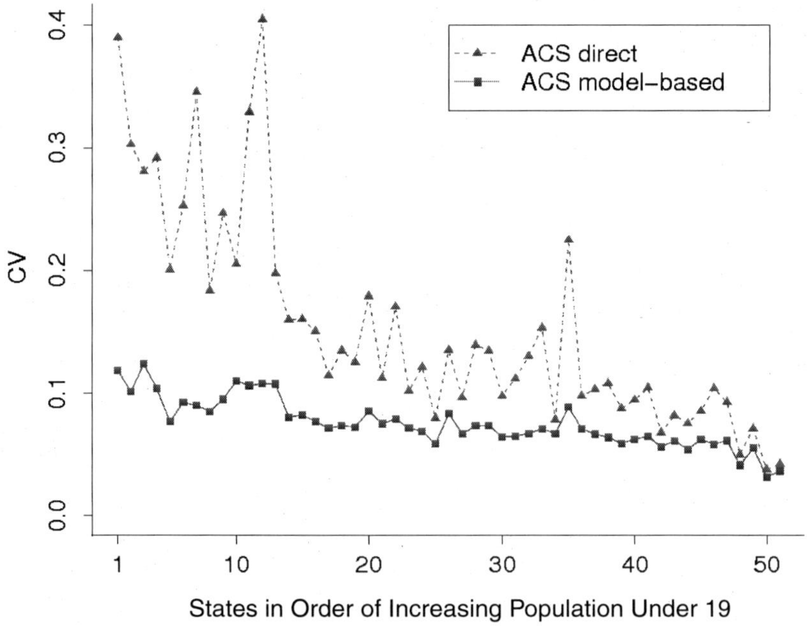

FIGURE 13-4 Estimated CVs by state for the ACS direct and the ACS model-based estimates of the number of uninsured children under age 19 with family income between 150 and less than or equal to 200 percent of poverty.
NOTE: States are ordered by increasing population under age 19.
SOURCES: Estimates are from the 2008 ACS survey and the 2008 ACS-based SAHIE model.

TABLE 13-1 Percentiles of Coefficients of Variation for SAHIE ACS Model-Based Estimates of the Number Uninsured and Corresponding ACS Direct Estimates

		Percentile					
Income		25	50	75	90	95	99
000-133	model	0.11	0.14	0.17	0.19	0.20	0.22
	direct	0.11	0.20	0.36	0.61	0.83	1.11
133-150	model	0.15	0.18	0.20	0.22	0.23	0.25
	direct	0.25	0.41	0.64	0.92	1.02	1.12
150-200	model	0.12	0.16	0.19	0.21	0.22	0.25
	direct	0.18	0.29	0.49	0.79	1.00	1.14
200-250	model	0.13	0.17	0.19	0.21	0.22	0.25
	direct	0.19	0.32	0.54	0.79	1.00	1.16
250-400	model	0.12	0.16	0.19	0.21	0.22	0.24
	direct	0.15	0.26	0.45	0.71	0.95	1.14
400+	model	0.15	0.18	0.20	0.22	0.23	0.25
	direct	0.17	0.32	0.55	0.85	1.01	1.17

NOTES: Observations for which ACS does not have a variance estimate have been dropped. There are approximately 1,500 state/age/race/sex estimates for each income group.
SOURCE: Estimates are from the 2008 ACS survey and the 2008 ACS-based SAHIE model.

ACS direct estimates. A somewhat common, if informal, criterion has been that estimates with CVs greater than .30 should be used with caution. In fact, the Census Bureau had adopted the standard that, for data to be released, the majority of key statistics should have CVs below 30 percent (U.S Census Bureau, 2007). The actual usefulness of estimates can be judged only in the context in which they will be used. Nevertheless, Table 13-1 shows that over 99 percent of the SAHIE estimates for these domains fall below that threshold.

FINAL REMARKS

This research shows that an ACS-based SAHIE model generally has lower measures of uncertainty compared with a CPS-based SAHIE model. This is not surprising. Even if the SAHIE program did not implement any changes in the model, the estimates would be more reliable and timely because 1 year of ACS data is needed as opposed to using a 3-year average of CPS ASEC data. The second goal of the paper was to see if the ACS survey data could support modeling more income-to-poverty ratios. For most small domains (state/age/race/sex estimates for each income group), ACS model-based estimates offer a large improvement over using survey-only estimates. With ACS direct estimates, many of the domains at the state-level have CVs that should be used with caution or not at all.

The ACS survey requires 5 years of data for estimates of all counties (U.S. Census Bureau, 2009). The first 5-year estimate of health insurance coverage will be the 2012 ACS to be released in fall 2013. Many of the small domains discussed in this paper should have a reliable 5-year health insurance coverage estimate at the state level. If a 5-year estimate of health insurance coverage is adequate for policy makers, the SAHIE state estimates should be viewed as a stopgap measure until 2013. It is unlikely that the ACS direct estimates will provide reliable county information for small domains (county/age/sex for each income group), even with multiyear data. It is likely that SAHIE county estimates for small domains will be reliable using underlying survey data of one year.

SAHIE estimates of uninsured people could be a valuable tool for policy makers to evaluate and administer means-tested programs, such as Medicaid and CHIP. These estimates can be used to target funds for outreach to specific uninsured and underserved demographic groups. States may want to use the estimates of the number of uninsured people, by income group, to approximate the total cost of subsidizing premiums for the health insurance exchanges.

FUTURE RESEARCH

Research on the SAHIE modeling is continuing. Some planned areas of research include the following:

- Using the ACS direct variance estimates. We can produce replicate-based estimates of the variances of the ACS estimates. In the past, we have not used such variance estimates. We have instead predicted variances of survey estimates within the model. We intend to investigate ways to use the information in the replicate-based variance estimates.
- Relaxing model assumptions. We intend to investigate alternatives to assumptions of independence and of constant variances.
- Predictors for insurance coverage. We have many predictors for income status but fewer that relate to insurance coverage that are not also linked to income status.
- Testing the model to incorporate more income groups (e.g., 250-300 and 300-400 percent of poverty) and age groups (e.g., 19-25-year-olds).
- Multiyear ACS estimates. One of the administrative sources is income data from the 2000 census. This income data will not be available from the 2010 census. We intend to investigate replacing the census data with the multiyear ACS.

ACKNOWLEDGMENT

This paper is released to inform interested parties of ongoing research and to encourage discussion of work in progress. Any views expressed on statistical, methodological, technical, or operational issues are those of the authors and not necessarily those of the U.S. Census Bureau.

REFERENCES

Bauder, M., Riesz, S., and Luery, D. (2008). *Further Developments in a Hierarchical Bayes Approach to Small Area Estimation of Health Insurance Coverage: State-Level Estimates for Demographic Groups*. Working Paper. Available: http://www.census.gov/did/www/sahie/publications/files/samb_jsm2008_paper_8.pdf [August 2010].

Blewett, L., and Davern, M. (2007). Distributing state children's health insurance funds: A critical review of the design and implementation of the funding formula. *Journal of Health Politics, Policy and Law, 32*(3), 415-455.

Centers for Medicare & Medicaid Services. (2010). *Medicaid Program—General Information—Technical Summary*. Available: http://www.cms.gov/MedicaidGenInfo/ [August 2010].

Czajka, J., and Jabine, T. (2002). Using survey data to allocate federal funds for the State Children Health Insurance Program (SCHIP). *Journal of Official Statistics, 18*(3), 409-427.

Dubay, L., Guyer, J., Mann, C., and Odeh, M. (2007). Medicaid at the ten-year anniversary of SCHIP: Looking back and moving forward. *Health Affairs, 26*(2), 370-381.

Fay, R., and Herriot, R. (1979). Estimates of income for small places: An application of James-Stein procedures to census data. *Journal of the American Statistical Association, 74*, 269-277.

Grumbach, K., Keane, D., and Bindman, A. (1993). Primary care and public emergency department overcrowding. *American Journal of Public Health, 83*(3), 372-378.

Kaiser. (2010). *Explaining Health Care Reform: Questions About Health Insurance Subsidies. Working Paper #7962-02*. Available: http://www.kff.org/healthreform/upload/7962-02.pdf [August 2010].

National Research Council. (2002). *Data Needs for the State Children's Health Insurance Program*. Panel for the Workshop on the State Children's Health Insurance Program. Editors A. Leibowitz and E. Pollack. Washington, DC: The National Academies Press.

Newacheck, P., Hughes, D., and Stoddard, J. (1996). Children's access to primary care: Differences by race, income, and insurance status. *Pediatrics, 97*(1), 26-32.

Newacheck, P., Hughes, D., Wong, S., and Stoddard, J. (2000). The unmet health needs of America's children. *Pediatrics, 105*(4), 989-997.

O'Hara, B. (2004). Do medical out-of-pocket expenses thrust families into poverty? *Journal of Health Care for the Poor and Underserved, 15*(1), 63-75.

O'Hara, B. (2008). Experimental health insurance estimates for low-income and demographic groups by state. *Health Services Research, 43*(5 Pt 1), 1693-1707.

O'Hara, B., Tangka, F. and Bauder, M. (in review). *New Health Insurance Estimates Improve Measuring Program Participation in the National Breast and Cervical Cancer Early Detection Program.*

P.L. 106-113. (1999). *Medicare, Medicaid, and SCHIP Balanced Budget Refinement Act of 1999.*

P.L. 111-3. (2009). *Children's Health Insurance Program Reauthorization Act of 2009.*

P.L. 111-148. (2010). *Patient Protection and Affordable Care Act of 2010.*

P.L. 111-152. (2010). *Health Care and Education Reconciliation Act of 2010.*

Small Area Health Insurance Estimates. (2009). *Model-Based Small Area Health Insurance Estimates (SAHIE) for Counties and States.* Working Paper. Available: http://www.census.gov//did/www/sahie/ [August 2010].

Small Area Health Insurance Estimates Team. (2008). *The Feasibility of Publishing County-level Estimates of the Number of Women Eligible for the CDC's NBCCEDP. Working Paper.* Available: http://www.census.gov/did/www/sahie/publications/files/cdc_feasibility_report_oct2008.pdf [August 2010].

U.S. Census Bureau. (2002). *Current Population Survey Design and Methodology Technical Paper 63rv.* Washington, DC: U.S. Government Printing Office.

U.S. Census Bureau. (2007). *Quality Requirements for Releasing Data Products.* Version 1.0. Washington, DC: U.S. Government Printing Office.

U.S. Census Bureau. (2009). *Design and Methodology American Community Survey.* Washington, DC: U.S. Government Printing Office.

Appendix A

Workshop Agenda and Participants

WORKSHOP AGENDA

Goals for the Planning Meeting:

1. Discuss the major issues associated with measuring health insurance coverage of children in the nation and states.
2. Hear from authors of commissioned papers on these issues.

Thursday, June 17, 2010

9:00 a.m. **Call to Order, Introductions, and Discussion of the Agenda for the Workshop**
V. Joseph Hotz, Chair

9:15 **Session 1: Congressional Interest in Estimates of Uninsured Children**
Chris L. Peterson, Congressional Research Service

9:45 **Session 2: Primer on Major Federal Surveys Used to Measure Health Insurance Coverage of Children (CPS, ACS, and NHIS)**
Michael Davern, NORC

10:15 ***Break***

10:30	**Session 3: Use of Administrative Data for Estimating the Coverage of Uninsured Children in States** *Michael Davern, Facilitator*
	A User's Perspective on the Quality of Administrative Data on Medicare and CHIP *David Rousseau, Kaiser Family Foundation*
	CMS Programs for Improving the State Administrative Microdata *David Baugh, CMS*
	Using Administrative and Survey Data for Programmatic Purposes *Richard Strauss, CMS*
11:45	*Working Lunch*
12:45 p.m.	**Session 4: State Use of Administrative and Survey Data: Exemplary Practices** *Joan Henneberry, Facilitator*
	The Massachusetts Experience: Using Data to Evaluate State Health Reform *Sharon Long, The Urban Institute*
	A Review of State Experience in Using Data to Evaluate CHIP Coverage *John McInerney, The Commonwealth Institute for Fiscal Analysis*
	Overview of State Survey Development and Use: What States Need to Monitor Health Insurance Coverage *Lynn A. Blewett, University of Minnesota and SHADAC*
2:30	*Break*
2:45	**Session 5: Modeling Strategies for Improving Estimates** *Eric V. Slud, Facilitator*
	Census Bureau Models for Measuring Health Insurance in Small Areas (SAHIE) *Brett O'Hara and Mark Bauder, U.S. Census Bureau*

Using Survey, Census, and Administrative Records Data in Small Area Models, with Illustrations from the Census Bureau's SAIPE Program
William Bell, U.S. Census Bureau

Joint Modeling of Survey and Administrative Record Data
Nathaniel Schenker, National Center for Health Statistics

4:45 p.m. *Adjourn*

Friday, June 18, 2010

9:00 a.m. **Session 6: Evaluation of the Adequacy of the Major Surveys to Measure Health Insurance Coverage for Children**
Lisa Dubay, Facilitator

Measuring Income
John Czajka, Mathematica Policy Research, Inc.

Measuring Health Insurance Coverage
Joanna Turner, SHADAC

Key Survey Characteristics: Frequency, Timeliness, Quality, Geographic Coverage
Genevieve Kenney, The Urban Institute

10:30 *Break*

10:45 **Session 7: The Future Need for Data on Children's Insurance Coverage**
Cindy Mann, Center for Medicaid and State Operations, CMS

11:30 **Summary: What Do We Need to Know?**
V. Joseph Hotz, Facilitator

Open Discussion Period

Adjourn

PARTICIPANT LIST

Committee Members

V. Joseph Hotz (*Chair*), Duke University
Lynn Blewett, University of Minnesota & SHADAC
Michael Davern, NORC at the University of Chicago
Lisa C. Dubay, Johns Hopkins University
Joan Henneberry, Colorado Department of Health Care Policy and Financing
Chris L. Peterson, Medicaid and CHIP Payment and Access Commission
Eric V. Slud, University of Maryland

Presenters

Mark Bauder, U.S. Census Bureau
David Baugh, Centers for Medicaid & Medicare Services
William Bell, U.S. Census Bureau
Cindy Mann, Center for Medicaid and State Operations
Brett O'Hara, U.S. Census Bureau
David Rousseau, Kaiser Family Foundation
Nat Schenker, National Center for Health Statistics, CDC
Richard Strauss, Centers for Medicaid & Medicare Services

Paper Authors

John Czajka, Mathematica Research Policy, Inc.
Jenny Kenney, The Urban Institute
Sharon Long, The Urban Institute
Victoria Lynch, The Urban Institute
John McInerney, The Commonwealth Institute for Fiscal Analysis
Joanna Turner, University of Minnesota & SHADAC

Guests

Eric Anderson, U.S. Government Accountability Office
Tere Angueira, U.S. Census Bureau
Jessica Bushar, NORC at the University of Chicago
Robin Cohen, National Center for Health Statistics
Steven Cohen, Agency for Healthcare Research and Quality
Jennifer Day, U.S. Census Bureau
Rick Denby, U.S. Census Bureau
Kathleen Farrell, National Academy for State Health Policy
Kenneth Finegold, HHS, Office of Health Policy
Bula Ghose, HHS, OS/ASPE
Brian Harris-Kojetin, Office of Management and Budget
Martha Heberlein, Georgetown University Health Policy Institute
Jacob Klerman, Abt Associates
Don Luery, U.S. Census Bureau
Lisa Mirel, National Center for Health Statistics, CDC
Shamis Mohamoud, NORC at the University of Chicago
Thomas D. Musco, HHS

Liz Pham, HHS, Office of Health Policy
Susan Queen, Health Resources and Services Administration
Donald W. Reese, HHS, Office of Health Policy
Mary C. Roary, OS/OMH Division of Policy and Data
Andrew Rock, HHS, ASPE
Thomas Selden, Agency for Healthcare Research and Quality

Sponsors

David S. Johnson, U.S. Census Bureau
Don Oellerich, Office of Human Services Policy, ASPE

NAS Staff

Connie Citro, Director, CNSTAT
Ben Galick, NAS Intern, DBASSE
Anthony Mann, Program Associate, CNSTAT
Tom Plewes, Study Director, CNSTAT
Gooloo Wunderlich, Senior Program Officer, CNSTAT

Appendix B

Biographical Sketches of Steering Committee Members

V. JOSEPH HOTZ (*Chair*) is arts and sciences professor of economics at Duke University. He is a research associate at the National Bureau of Economic Research and the Duke Population Research Institute and a research affiliate of the Institute for Research on Poverty and the National Poverty Center. He is also a fellow of the Econometric Society. Previously, he served as professor and chair of the Department of Economics at the University of California, Los Angeles. His research interests include labor economics, economic demography, and evaluation of the impact of social programs. He has served on several panels of the Committee on National Statistics, including those on the Census Bureau's re-engineered SIPP, and access to research data: balancing risks and opportunities. He has a Ph.D. degree in economics from the University of Wisconsin.

LYNN A. BLEWETT is principal investigator and director of the State Health Access Data Assistance Center (SHADAC) and professor in the Division of Health Policy and Management at the University of Minnesota School of Public Health. She also directs the State Health Access Reform Evaluation (SHARE), a national program office to fund and synthesize rigorous evaluation of state health reform initiatives. Her current area of research is defining and measuring underinsurance and documenting the development of local access to care programs, community-based initiatives to meet the primary health care needs of the growing number of uninsured adults. Her prior health policy experience includes work for the U.S. Senate and the Minnesota Department of Health. She has a

Ph.D. degree in health services research, policy, and administration from the University of Minnesota, a M.A. degree in public affairs from the Hubert H. Humphrey Institute of Public Affairs at the University of Minnesota, and a B.A. degree in psychology from the University of Wisconsin, Madison.

MICHAEL DAVERN is vice president and director of the Public Health Research Department in the National Opinion Research Center (NORC) at the University of Chicago. Prior to this appointment, he was at the University of Minnesota School of Public Health, where he was an assistant professor of health policy and management and research director of the State Health Access Data Assistance Center (SHADAC). He also codirected the U.S. Census Bureau Research Data Center at the university. He previously served as a statistician for the Labor Force and Transfer Programs Statistics Branch of the U.S. Census Bureau. A major focus of his work has involved applying state-level data to health policy issues and helping states monitor trends in health insurance coverage rates. He has a Ph.D. degree in sociology from the University of Notre Dame, a M.A. degree in sociology from Colorado State University, and a B.A. degree in sociology from Saint John's University.

LISA C. DUBAY is a health services researcher with over 20 years experience in public health. She recently joined the faculty at the Johns Hopkins Bloomberg School of Public Health as associate professor. Prior to this, she was a principal research associate at the Urban Institute, where she focused on the effects of public policies on insurance coverage, access to care, and health outcomes for low-income populations. She has led numerous national evaluations of public expansions in coverage for federal agencies and private foundations. Her dissertation examined socioeconomic gradients and racial disparities in children's health status and well-being and their emotional and cognitive functioning. She has extensive experience analyzing nationally representative household surveys. She has Ph.D. and S.C.M. degrees from the Johns Hopkins University.

JOAN HENNEBERRY is executive director of the Colorado Department of Health Care Policy and Financing and leads the agency responsible for managing public health insurance programs, including Medicaid and Child Health Plan Plus. She is a senior health policy adviser to the governor, developing and implementing health reform policies and initiatives. Prior to the cabinet appointment, she worked in the private sector after spending 7 years at the National Governors Association providing consultation to states on health care services and financing, cost containment, and emerging policy issues. She spent 13 years at the Colorado

Department of Public Health and Environment, chairs the board of the Colorado Regional Health Information Organization, and serves on the Executive Committee for the National Academy for State Health Policy. She has a M.A. degree in management, and completed the Senior Executives in State and Local Government Program at the Kennedy School of Government at Harvard University.

CHRIS L. PETERSON is director of eligibility, enrollment, and benefits at the new congressional commission MACPAC, the Medicaid and CHIP Payment and Access Commission. Prior to this, he was specialist in health care financing in the Domestic Social Policy Division of the Congressional Research Service (CRS). He has written and testified extensively on Children's Health Insurance Program (CHIP) funding issues and formula allocations, as well as on survey estimates of low-income uninsured children. His work for CRS also covered issues and options for a health insurance "exchange," risk pooling, mandates, minimum benefit standards, actuarial values, subsidies, tax policy, public programs, and other federal statutes and regulations. Previously, he worked at the Agency for Healthcare Research and Quality and the National Bipartisan Commission on the Future of Medicare (Breaux-Thomas). He has an M.P.P. degree from Georgetown University's Graduate Public Policy Institute and a B.Sc. degree from Missouri Western State University.

ERIC V. SLUD is professor in the statistics program at the University of Maryland. His research includes the evaluation of models for small-area poverty estimates from census and Current Population Survey (CPS) data and analysis of nonresponse-adjusted survey estimates. He is a fellow of the American Statistical Association and a consultant to the U.S. Census Bureau. He was a member of the National Research Council (NRC) committee to review K-12 standards in mathematics and served as a reviewer for a previous NRC study on the American Community Survey. He has a B.A. degree from Harvard College and a Ph.D. degree in mathematics from the Massachusetts Institute of Technology.

COMMITTEE ON NATIONAL STATISTICS

The Committee on National Statistics (CNSTAT) was established in 1972 at the National Academies to improve the statistical methods and information on which public policy decisions are based. The committee carries out studies, workshops, and other activities to foster better measures and fuller understanding of the economy, the environment, public health, crime, education, immigration, poverty, welfare, and other public policy issues. It also evaluates ongoing statistical programs and tracks the statistical policy and coordinating activities of the federal government, serving a unique role at the intersection of statistics and public policy. The committee's work is supported by a consortium of federal agencies through a National Science Foundation grant.